INVASIVE PLANTS OF THE UPPER MIDWEST

Elizabeth J. Czarapata

Invasive
Plants
of the Upper Midwest

AN ILLUSTRATED GUIDE TO

THEIR IDENTIFICATION

AND CONTROL

THE UNIVERSITY OF WISCONSIN PRESS

This book was published with the generous financial support of
The Park People, Friends of Milwaukee County Parks
The Endangered Resources Program of the Wisconsin Department of
     Natural Resources
The America Outdoors Program of the United States Bureau of
     Land Management
The Bradshaw-Knight Foundation
and many friends concerned about the environment.

The University of Wisconsin Press
1930 Monroe Street
Madison, Wisconsin 53711

www.wisc.edu/wisconsinpress/

3 Henrietta Street
London WC2E 8LU, England

5   4   3

Text design by Mira Nenonen
Printed in Canada

Library of Congress Cataloging-in-Publication Data
Czarapata, Elizabeth J.
Invasive plants of the upper Midwest: an illustrated guide to their identification and
control / Elizabeth J. Czarapata.
   p. cm.
   Includes bibliographical references and index.
   ISBN 0-299-21054-5 (pbk.: alk. paper)
   1. Invasive plants—Middle West—Identification. 2. Invasive plants—Middle West—
Pictorial works. 3. Invasive plants—Control—Middle West. I. Title.
SB612.M54C93   2005
581.6'2'0977   2004025635

Dedicated to my husband, Lee,
for his incredible patience and support.

Contents

Foreword

THIS BOOK EMERGED from Betty Czarapata's desire to educate her friends, neighbors, students, and others about the ecological hazards posed by garlic mustard, common buckthorn, and the other species that she observed replacing the native vegetation in the parks and other wild places she loved in southeastern Wisconsin. Frustrated at not being able to identify these invasive plants, Betty began taking photographs in order to study them further. She soon began using her photographs in talks and displays, and found that others were just as eager to learn about these plants.

At the same time, several state natural resource agencies had developed plant control manuals, but none had good photos to help people identify the plants in all of their life stages. Betty set out to change that. In 1999 she self-published *Invading Weeds: A Growing Threat to Wisconsin's Biodiversity*—a compilation of photographs and species information about invasive plants in Wisconsin. Then she set her sights higher and began to focus on a similar, but more detailed, guide for the entire region. What began as a simple quest for an educational tool grew into the most comprehensive guide to date on the identification and control of invasive plants in the upper Midwest.

Betty envisioned this book as a guide to help both the novice gardener and the experienced land manager identify and contain the plants that invade the natural areas of the upper Midwest. She worked diligently at this task scouring the literature and consulting with experts to find the latest and most complete information. I think the results of her efforts are outstanding. The habitat and range information in this book will help people know which species to look for on their land. The numerous photos and detailed descriptions allow the reader to identify invasive plants in all their various life stages. Property owners and land managers will find extensive information about control techniques for each species and type of site. Chapter 2 goes into detail about the various control methods, including guidelines for the safe and effective use of herbicides. Chapter 7 explains what property owners and land managers can do to restore a site to native species once the invasive plants are removed. The numerous appendices provide references to check for more detailed research, and lists of current websites to stay up-to-date with new information and ongoing research. An extensive glossary provides definitions for the terms used in the plant descriptions. A matrix of all the plants listed in the book, and all other species known or suspected to have the potential to become invasive in the region, is also included, along with basic information such as habitat and range invaded and what controls can be used.

In the course of compiling this work, Betty was diagnosed with ovarian cancer. Between her chemotherapy treatments she labored on the book, writing and rewriting the species accounts, selecting the best photos out of her large collection, and gathering the most current information available on control methods. She sought out specialists to review various sections of the book in an attempt to make it as accurate as possible. By mid-December 2003 she had completed all the parts that were critical for her to finish. Unfortunately for all of us, especially those who knew her and worked with her, she died a few days after Christmas that year. She knew that her efforts would be successful, however, and the book would be carried on without her.

Our native landscapes have evolved over the millennia, gradually shifting with various environmental and human influences. But the destructive transformation that has been wrought on the landscapes of North America by the combination of human-induced disturbance and non-native invasive plants has come about so quickly that our remaining natural areas are in danger of being seriously degraded.

Betty Czarapata was never able to hold this book in her hands, but you are. It was her hope that you would become familiar with the plants in this book and do what you can to help keep our wild areas thriving with a diversity of native species. The upper Midwest still has a fair number of forests, lakes, and other habitats not yet overrun by invasive plants. With cooperation from you and many other like-minded people, we can keep some of these areas free of invasives and restore their natural biodiversity.

KELLY KEARNS

Minnesota

Wisconsin

Michigan

Iowa

Illinois

Indiana

Ohio

Missouri

About This Book

IT WOULD BE IMPOSSIBLE for this book to discuss in detail every plant that is potentially invasive in the upper Midwest. In order to include information about a wider selection of species, an invasive plant list, along with general plant characteristics, can be found in Appendix E.

Plants in chapters 3, 4, 5, and 6 are arranged according to their general growth pattern (Trees and Shrubs, Vines, Forbs, Grasses and Grasslike Species, and Aquatic Plants), then alphabetically according to their scientific names. Although some plants discussed in the Trees sections of the chapters are sometimes considered shrubs and develop a more compact habit under certain growing conditions, they are classified as trees here due to their general structure and most common features.

A plant may also have acquired more than one scientific name over time as a result of being moved from one genus to another, or because it was officially described and recorded more than once. Plants are listed here according to the scientific name provided by Gleason and Cronquist's *Manual of Vascular Plants of Northeastern United States and Adjacent Canada,* second edition (1991). If another scientific name for the same plant is frequently used, it will be preceded by "syn." (synonym).

If more than one species in the same genus are discussed together, such as *Lonicera tartarica* and *Lonicera morrowii,* the second species discussed will usually have the genus name (*Lonicera* in this case) abbreviated to the first letter, as in *L. morrowii.*

If you would like to find information on a particular plant and know a common name for it (plants often have several common names) but do not know its scientific name, check the index at the back of the book to see if the common name is listed and the corresponding page where it is discussed. Plants with Native American-derived common names are often indigenous to North America, suggesting that white settlers had not encountered them in Europe.

Readers are encouraged to become familiar with the species discussed in this book, especially those in chapter 3, "Invasive Plants of Major Concern," and they should monitor their land for the presence of these species and begin control efforts where feasible. Preserving native plant diversity should be the goal.

Before attempting to eliminate any weed colony, with or without the use of herbicides, read chapter 2, "Invasive Plant Control Techniques," for general herbicide information and practical weed control suggestions.

Acknowledgments

Wisconsin Manual of Control Recommendations for Ecologically Invasive Plants

I must express my sincere gratitude to the Bureau of Endangered Resources of the Wisconsin Department of Natural Resources, and to Randy Hoffman and Kelly Kearns in particular. Without their information and guidance, this book would not have been written. Much of the information used to compile *Invasive Plants of the Upper Midwest* was obtained from the *Wisconsin Manual of Control Recommendations for Ecologically Invasive Plants,* which Randy and Kelly coordinated and edited. Because their manual and other work have been extremely valuable in moving this educational effort forward, it is fitting to acknowledge all those who participated in its development.

The first draft of the manual (1993) was coordinated and edited by Randy Hoffman with assistance from Paul Matthiae, Mark Martin, Jim Reinartz, the late Virginia Kline, Rich Henderson, Steve Glass, Wayne Pauly, Nancy Braker, and Kelly Kearns.

Information for the second version (1997) was provided by all of the original contributors as well as Victoria Nuzzo, Steve Richter, Bill Kleiman, Jim Sime, Gary Eldred, Russ Hefty, JoAnn Cruse, Dan Spuhler, Brian Swingle, Ed Bergman, Brock Woods, Peggy Chung, Beth Goeppinger, Alice Thompson, Joanne Kline, David Kopitzke, and Steve Apfelbaum. Several of the above mentioned also assisted in editing the manual. Student assistants Quan Bahn, Jamie MacAlister, Irene Sedowski, David Callery, and Elizabeth Hermsen, under the direction of Kelly Kearns, helped with literature searches, writing, editing, and compilation of the 1997 version.

Additional Acknowledgments

Many other individuals have helped in a variety of ways to make this book possible. They include: Bob Wakeman, Frank Koshere, Laura Herman, Scott Provost, Amy DeMars Staffen, Scott Toshner, Matthew L. Burczyk, and Rachel Armstrong, Wisconsin Department of Natural Resources; Evelyn A. Howell, UW-Madison Landscape Architecture Department; Dave Egan and Steve Glass, UW-Madison Arboretum; Jerry Doll, Ph.D, University of Wisconsin Agronomy Department; Rick Schulte, UAP Timberland; Jerry Schwarzmeier, Retzer Nature Center; Ken Solis, MD, The Park People of Milwaukee County, Inc.; Theodore Cochrane, UW-Madison Department of Botany; Neil Luebke, Milwaukee Public Museum; Brian Swingle, Anne Parrish, Ph.D, and Ed Bergman, Ph.D, Wisconsin Department of Agriculture, Trade, and Consumer Protection; Richard Barloga; Larry Leitner, Southeast Wisconsin Regional Planning Commission; Laurie Weiss, University of Wisconsin-Extension, Milwaukee; Dick Stark and Karla Leithoff, Wisconsin Department of Transportation; Steve Apfelbaum and Aaron Kubichka, Applied Ecological Services; George Kitt, United States

Environmental Protection Agency; Larry Lamb, University of Waterloo; Randy Powers, Prairie Future Seed Company; Ginny Haen, Whitefish Dunes State Park; John and Martha Schwegman; Jody Shimp, Illinois Department of Natural Resources; Adrian Barta, Wisconsin Department of Agriculture; Beth Shimp, National Forest Service; Ann (Mackenzie) Grewe; Carol Owens; Bill McClain, Illinois Department of Natural Resources; Mariette Nowak and Mark Verhagen, Wehr Nature Center; Jil Swearingen, National Park Service, Center for Urban Ecology; Jennifer L. Windus, Ohio Department of Natural Resources; Dr. John and Leah Schaut, S. Galen Smith, Ph.D.; and Bob Retko, Bradley Sculpture Garden.

Scientific Plant Names

Where applicable, scientific names for plants discussed in this book were obtained from the *Manual of Vascular Plants of Northeastern United States and Adjacent Canada*, 2nd ed., Henry A. Gleason, Ph.D, and Arthur Cronquist, Ph.D., 1991. New York Botanical Garden, Bronx, NY.

A second source for scientific and common names was *A Synchronized Checklist of the Vascular Flora of the United States, Canada, and Greenland*, 2nd ed. J.T. Kartesz. 1994. Timber Press, Portland, OR.

Photography

Elizabeth J. Czarapata took all the photographs in this book except for those indicated immediately below. All other photographs are used by permission.

Additional photos by: Jil Swearingen, National Park Service, Center for Urban Ecology (lesser celandine and mile-a-minute vine), Bill McClain, Illinois Department of Natural Resources (kudzu), Thomas D. Brock (cut-leaved coneflower), Robert W. Freckmann, University of Wisconsin-Stevens Point (giant lupine), Emmet Judziewicz, University of Wisconsin-Stevens Point (giant lupine, queen of the meadow), Dave Egan (Norway maple, Eastern red cedar, and switchgrass).

Matthew Burczyk scanned and compiled all the photos, and set them in order for the book designer.

Illustrations

Plant sketches are from *Aquatic and Wetland Plants of Northeastern North American*, Volumes 1 and 2, by Garrett E. Crow and C. Barre Hellquist; and *Through the Looking Glass: A Field Guide to Aquatic Plants* by Susan Borman, Robert Korth, and Jo Temte, with illustrations by Carol Watkins. Permission was obtained for the use of these illustrations. I also want to thank the people who provided illustrations that we did not use.

Production Credits

Final editing by Dave Egan. Preliminary editing in part by Kelly Kearns, Jerry Doll, Rick Schulte, Matthew L. Burczyk, Lee Czarapata, Ken Solis, MD, Laura Herman, Galen Smith, Scott Provost, Amy DeMars Staffen, Frank Koshere, and Kathy Winkler.

Graphic designer: Graphic Composition, Inc., with contributions by Linda Pohlod and Georgene Schreiner.

Additional assistance by Susan Pittelman of Printstar Books and Ann (Mackenzie) Grewe.

Special recognition

Special recognition is extended to The Park People of Milwaukee County, Inc., Wild Ones Natural Landscapers, the Milwaukee Audubon Society, and all those who supported me in a multitude of ways at a very difficult time. My heartfelt thanks to everyone involved.

Introduction

"GARLIC MUSTARD"—What an enticing name! That's what I thought when I first heard it mentioned several years ago. In casual conversation with a new friend, I was warned about this European species that, according to her, was causing enormous damage in many Wisconsin woodlands by crowding out native wildflowers.

Although I had worked on numerous environmental issues as a volunteer, trying to familiarize myself with everything from erosion control to global warming, the topic of garlic mustard or invasive weeds had never come up. Like countless other Midwestern citizens, this was one environmental problem that I was truly in the dark about.

But, my curiosity was stirred. As I tried to learn more about these plants and their disastrous ability to take over natural areas to the detriment or exclusion of other species, the Bureau of Endangered Resources (BER) from the Wisconsin Department of Natural Resources assisted me by providing the names, identifying characteristics, sketches, and available information about the most troublesome plants. Slide presentations by Kelly Kearns of the BER and others helped give me some basic knowledge. Although very unsure of myself, I set out to find these deceptive and sometimes beautiful troublemakers.

Unfortunately, I soon discovered that positively identifying invasive weeds, with my typical suburban-homeowner knowledge about natural areas, was next to impossible. There never seemed to be a knowledgeable person handy when I needed to know if the plant I was observing was considered ecologically invasive. So, I began to take photographs of suspected invaders, later having them identified by experts in Wisconsin vegetation. Thus began the photo collection that eventually resulted in this book.

As I gradually began to recognize the worst invasive weeds, my curiosity about them changed to alarm. Many high-quality woodlands, often comprised of 120 or more native plant species, were being quietly and tragically transformed into haunting Eurasian jungles of buckthorn, honeysuckle, and garlic mustard. Spring wildflowers, the "Mona Lisas" of our woodlands that delight so many, are being lost in the process still today. In addition, many woodland understories are so filled with invasive weeds that tree saplings are being shaded out, rendering them unable to replace the mighty oaks, hickories, and maples when they eventually die. Countless other native trees and shrubs are also at risk. To lose these vital components of Midwestern woodlands is unthinkable. Similar stories can be told about our prairies, wetlands, lakes, and other aquatic areas. As native plants disappear, many forms of wildlife that evolved with them and often rely on them for survival will also disappear. Invasive weeds are indeed a *real* and ever-growing problem.

The good news is that this is one environmental problem that we can do something about. I have seen the tremendous difference that even a few individuals can make in the battle to regain the land for native species. What a sense of accomplishment when people realize that a trillium, shooting star, or cardinal

flower can go on living due to their weed control efforts and that future generations may now inherit more than a weed patch!

This book will, I hope, help you become better acquainted not only with invasive weeds but also with natural areas in general. Ultimately, your appreciation
for our natural heritage will grow, you will remove invasive weeds from your
land, plant more locally native species, and seek appropriate opportunities to become a steward of natural areas beyond your own property lines.

Mel Ellis once said in *Notes from Little Lakes,* "The only way to form a true
partnership with nature is if we invest sweat, blisters and aching backs. Then,
most stones have felt our touch, and most blooms, seeds and fruits are the harvest of our efforts. So, like a loving mother, we care. . . . No disappointments can
permanently tarnish our love affair. We celebrate anniversary after anniversary
as acolytes at spring's priestly ritual of re-creation."

Recent generations have sadly lost touch with nature. There is no better time
than now to venture out into the *real* world and get reacquainted. A truly grati-
fying experience awaits you!

Happy weeding!

Elizabeth J. Czarapata

INVASIVE PLANTS OF THE UPPER MIDWEST

Why Invasive Plants Cannot Be Ignored

SLOWLY, BUT PERSISTENTLY, making their way across the land, ecologically invasive plants are the silent invaders of our times. They destroy three million acres each year in the United States, cost our society $35 billion annually, devour even our finest natural areas, and pose a major threat to restored areas and endangered species. Invasive plants affect us all. Farmers, suburban homeowners, and apartment dwellers face increasing costs for food and lumber production, while property values and recreational opportunities decline in areas where invasive species have taken hold. The beauty and biological diversity associated with healthy natural areas are being lost as well.

As international trade expands and the human population continues to rise, the spread of ecologically invasive plants takes its toll. A tragedy is relentlessly unfolding—yet few people recognize these weeds or are aware of the consequences of allowing them to proliferate. To the untrained eye, the lush, green landscapes often associated with invasive plants create illusions of vibrant, flourishing ecosystems when, in fact, many species have been lost and complex natural processes have been disrupted. Greater understanding of the fragile and amazing natural world that surrounds and sustains us is desperately needed.

A 1993 report by the U.S. Congress Office of Technology Assessment titled, *Harmful Non-Indigenous Species in the United States,*[1] predicted that if we continue with "business as usual," biological invasions will become one of the most prominent ecological issues on earth by the middle of the twenty-first century. Invasive organisms, such as gypsy moths, Asian long-horned beetles, zebra mussels, fire ants, snake-head fish, and the West Nile virus, have received significant public attention in recent years—and deservedly so. This book, however, will focus specifically on invasive plants that affect the prairies, forests, wetlands, lakes, and other native plant communities of the upper Midwest.

WHAT ARE INVASIVE PLANTS?

The Invasive Plant Association of Wisconsin defines invasive plants as "non-indigenous species or strains that become established in natural plant communities and wild areas and replace native vegetation." Executive Order 13112, signed by President Clinton in 1999 to address the issue of invasive species, further states that they are "likely to cause economic or environmental harm or harm to human health." "Non-indigenous" species are also known as "non-native," "alien," or "exotic" species. These terms, for the purposes of this book, will refer to species that did not occur in a specific area or plant community before the European settlement of the upper Midwest.

Invasive species have earned official standing as a leading threat—second only to habitat destruction—to the native species of the United States. (William Stolzenburg, Science editor, *Nature Conservancy,* July–August 1999)

Numerous studies show reduced numbers of birds, reptiles, small mammals and insects in stands of nonnative plant species. (Jerry Asher, "War on Weeds: Winning it for Wildlife," Bureau of Land Management, March 2000)

[Invasive] weeds are spreading at about 4,600 acres per day on Western federal lands alone (outside of Alaska). (Jerry Asher, "War on Weeds: Winning It for Wildlife," Bureau of Land Management, 2000)

Biological diversity is our most valuable but least appreciated resource. (E. O. Wilson, University Research Professor and Honorary Curator in Entomology, Harvard University, Cambridge, MA)

Invaders don't simply consume or compete with native species—they change the rules of existence for all species by altering ecosystem processes such as primary productivity, decomposition, hydrology, geomorphology, nutrient cycling or natural disturbance regimes. (Peter M. Vitousek, Carla M. D'Antonio, Lloyd L. Loope, and Randy Westbrooks, *American Scientist,* 84[5], September–October 1996)

WHERE DO INVASIVE PLANTS COME FROM?

Nearly any place in the world can produce plants that can become invasive elsewhere. Such plants typically become disruptive in regions that have a somewhat similar latitude and climate to their native range. Because they often occupy large north–south areas of their homeland and are adapted to a wide variety of growing conditions, their impact is likely be felt over wide geographic areas here. The majority of plants that have become troublesome in the upper Midwest are from Europe and Asia.

WHY ARE INVASIVE PLANTS A PROBLEM?

Invasive plants are a problem because they have competitive advantages over native plant species. These advantages include:

- The absence of plant diseases, insects, and other plant-eating organisms that help keep plant populations in check in their place of origin.
- The ability to grow and leaf-out earlier than native plants, which gives invasive plants a competitive advantage in terms of capturing sunlight, soil moisture, and soil nutrients—all essential components of healthy plant growth.
- The ability to reproduce both rapidly and in great numbers, often from seed or other plant parts that remain viable in the ground for long periods of time.
- The ability to thrive in a wide variety of habitats and soil conditions.

A plant with many or all of the above characteristics is likely to have a significant competitive advantage over many native plant species.

ARE ALL NON-NATIVE PLANTS A PROBLEM?

The vast majority of non-native plants cause little, if any, environmental damage other than, perhaps, taking up space that could be occupied by native species. Our food supply, for example, is comprised of many nonindigenous species. Many non-indigenous plants also provide medicinal or other benefits. They occupy their place in the landscape and pose little or no threat to natural areas. Many cannot even survive outside of cultivation.

But some non-indigenous plants, about 15 percent, are not so benign. Once they take hold outside their natural range, these invasive species begin to reproduce abundantly and cause significant environmental damage.

WHAT IS THE DIFFERENCE BETWEEN AGRICULTURAL WEEDS AND INVASIVE PLANTS?

Most, but not all, agricultural weeds are non-indigenous to the areas they currently inhabit. Agricultural weeds are primarily disturbance-oriented species that are managed with various conventional manual, mechanical, and chemical control methods. Agricultural weeds in the upper Midwest tend to be annual or biennial species that are either herbaceous or grasses. Agricultural weeds don't tend to be big problems in natural areas, although they may be in ill-prepared ecological restorations on former agricultural land. In addition to herbaceous and graminoid species, natural area land managers, unlike their agricultural counterparts, must contend with woody species. Farmers control agricultural weeds by plowing and herbiciding, while natural area managers use a variety of control techniques including prescribed burning, herbicides, and manual weeding.

How Did Invasive Plants Get Here?

The introduction of invasive plants to the upper Midwest and other parts of the world has generally resulted from intentional or accidental human activities. These include:

- The sale and distribution of ornamental plants and seeds. More than 150 non-native plants that were originally introduced as ornamentals have become invasive in natural areas in the United States.[2]
- Planting certain species for agricultural purposes, fiber production, wildlife habitat, or erosion.
- The arrival of seeds or other propagules of aquatic plants in the ballasts of ships, holds of boats, and aquarium stock.
- The use of weed-contaminated agricultural or nursery seed.
- The transfer of seeds or other propagules in soil and nursery stock, or with livestock.

Once a plant begins to reproduce outside its natural range, its dissemination is often enhanced by such things as:

- Wind and water movement, including flooding.
- Seeds sticking to clothing, animal fur, or to mud in the soles of shoes or in the treads of tires.
- Agricultural machinery, railroad cars, earth-moving equipment, mowers, and the like.
- Moving hay or other forage crops, sometimes contaminated with weeds.
- Birds and other animals eating fruits and berries, and then moving about the landscape excreting the seeds.
- Using dried plants in craft arrangements.
- Intentional planting of showy, but often invasive, non-native wildflowers.

Why Should We Be Concerned About Invasive Plants?

Invasive plants have a staggering impact on our environment and quality of life. Allowing them to spread unchecked has many consequences of ecological, economic, and aesthetic/historical significance. There are also ethical issues to consider.

The total number of harmful [non-indigenous species] and their cumulative impacts are creating a growing burden for the country. We cannot completely stop the tide of new harmful introductions. Perfect screening, detection, and control are technically impossible and will remain so for the foreseeable future. Nevertheless, the federal and state policies designed to protect us from the worst species are not safeguarding our national interests in important areas.

These conclusions have a number of policy implications.

First, the nation has no real national policy on harmful introductions; the current system is piecemeal, lacking adequate rigor and comprehensiveness.

Second, many federal and state statutes, regulations, and programs are not keeping pace with new and spreading non-indigenous pests.

Third, better environmental education and greater accountability for actions that cause harm could prevent some problems.

Finally, faster response and more adequate funding could limit the impact of those that slip through. (U.S. Congress, Office of Technology Assessment, *Harmful Non-Indigenous Species in the United States,* September 1993)

Ecological Consequences

- Biological diversity—the vast array of plants, animals, and microbes in any given area—may decline.
- Native plants may be crowded or shaded out of existence. Some invasive plants produce chemicals that inhibit the growth of other plant species. As a result, wildlife habitat and food sources are degraded or destroyed. Wildlife often rely on specific native plants to survive. If the diversity and quantity of native plants diminish, the diversity and quantity of native wildlife may diminish. Local or widespread extinction of some species may result.
- Threatened and endangered species—those species that are particularly vulnerable to environmental disruption—face more rapid decline once their habitat is infested with invasive weeds.
- Hybridization between invasive and native species can sometimes occur, leading to the potential loss of the original native strains.
- Soil instability and runoff may increase due to the loss of groundlayer vegetation.
- The need to use herbicides tends to increase the longer invasive plants are ignored.
- Flooding and fire regimes, microbial activity, soil characteristics, and other natural processes can be altered.
- Water quality and water quantity may decrease in a given area.
- Plant pests or diseases may threaten desirable plants in areas where invasive plants act as their hosts. Common buckthorn, for example, is an alternate host for the soybean aphid.

Economic Consequences

- Invasive plants cost the United States economy more than $35 billion a year.[3]
- Invasive trees and shrubs can significantly decrease the regeneration of trees needed for food and fiber. Many of our forests are turning into thickets of unusable woody weeds. This change is gradual, but foresters are seeing the potential for serious impacts on future forest production.
- The presence of invasive plants in some areas can greatly reduce property values.
- Invasive plants are a major threat to recreational uses of land and water (hunting and fishing, swimming, hiking, photography, birding). They are often thorny, scratchy, poisonous, or simply too dense to get through, limiting access to recreational and other areas.
- Some invasive plants cause human and livestock health concerns, such as the skin burns caused by wild parsnip and poisonous compounds produced by other plants.
- Future products may never be developed if invasive plants cause the loss of plants with undiscovered useful properties.

Aesthetic/Historical Consequences

As populations of invasive plants increase, the seasonal beauty and natural heritage that native plants have given us for generations are at risk of being lost. Imagine a spring without native wildflowers, a forest without oaks, a fall without nature's splendid display of color, or a wetland without wildlife. It has taken ten thousand years, since the retreat of the last glaciers, for our native plant communities to evolve to their present natural state. Our understanding of how nature's intricate web of life sustains itself is far from complete. While we can replant native species and restore many facets of the natural world around us, completely duplicating what once was is impossible. In short, if invasive plants are allowed to destroy our remaining natural areas, their pristine qualities and abilities to teach us what we do not yet know will essentially be gone forever.

ETHICAL IMPLICATIONS

We have an ethical obligation to pass along a world that is at least as healthy as the one we now enjoy. Land management decisions today can affect all future generations. To do nothing and neglect our wild lands will lead to their eventual demise. Although everyone is limited in the amount of muscle, time, or money they can devote to invasive plant control, the majority of Americans need to be made aware of invasive plants and take some action. Whether it is by eliminating invasive plants from their own yards, removing them from watercraft, or simply bringing them to the attention of neighbors and public officials, we all share some responsibility to help maintain the health of our natural areas. If we want to make sure that our descendants are able to gaze upon a prairie in bloom, hear the enchanted chorus of songbirds in spring, swim in a nearby pond, or use any plant now in existence for medicinal purposes, we cannot ignore the pernicious, escalating spread of invading plants!

> Weed management efforts have a higher probability of success when adjacent landowners, public land users, agencies, universities, or other interested people are participating. (Jerry Asher, "War on Weeds: Winning It for Wildlife," Bureau of Land Management, 2000)

KEN SOLIS
"Convincing the Skeptics"

KEN SOLIS, A PHYSICIAN with considerable knowledge and concern about natural areas, was instrumental in helping The Park People of Milwaukee County, Inc. create the Weed-Out Program in 1996. The Weed-Out Program is designed to educate the public about ecologically invasive weeds and encourage them to become involved in local weed-control efforts. Hundreds of volunteers participate in this program, and many important natural areas in the Milwaukee County Park System have been saved since the program began.

Although The Park People have made significant progress in terms of raising public awareness about the adverse effects of invasive plants, some people still need to be convinced that controlling invasive plants is important. With this in mind, Ken has put some thoughts together to help land managers, property owners, concerned citizens, and volunteers who may have to deal with skeptics.

WHY YOU MIGHT HAVE TO DEAL WITH SKEPTICS

- To convince neighbors and political entities about the importance of ecological restoration efforts, including the removal of invasive plants.
- To answer criticism or generate publicity in a public forum, including newspapers, radio, or television.
- To persuade your friends or acquaintances.
- To help in fund-raising efforts and grant writing.

GENERAL GUIDELINES

- Avoid emotional confrontation.
- Educate relevant political entities and households adjacent to the restoration site before culling, planting, or otherwise making changes to an area.
- Acknowledge the perceptions of politicians and neighbors and be prepared to consider compromises.
- Be well prepared—know the critical facts that will answer their concerns.
- Have educational material, such as brochures and flyers, on hand. Educational material from a respected third party is especially effective.
- Form alliances with other environmentally oriented groups, individuals, and policy makers.
- Recognize that some people have a hard time rationalizing controlling ecologically invasive plants because the plants they know don't cause problems in the manicured urban/suburban landscape, but instead are "out there" in the nature preserve, wilderness, or on the roadside.
- Recognize that the more you visually change an area, the harder you will have to work to convince people you are doing the right thing.

ARGUMENTS SKEPTICS MAKE AGAINST INVASIVE PLANT CONTROL AND RESPONSIBLE ANSWERS TO THEIR POSITIONS.

Argument: Invasive plants are created by God or Nature and, therefore, belong where they are.

Rebuttal: Using that argument, we shouldn't try to control diseases like influenza or malaria, invasive animals like fire ants and Africanized bees, or noxious plants like poison ivy and ragweed.

Argument: During the history of Earth, species have crossed ocean and mountain barriers many times. Nothing really new is happening now.

Rebuttal: What has dramatically changed in modern times is the magnitude and pattern of non-native species introductions around the globe due to human activity. The Hawaiian Islands, the most isolated island group in the world, provide a model of how humans affect ecosystems through the introduction of new species.

Before the arrival of humans, scientists estimate that a new species would find its way to the islands of Hawaii about once every ten thousand years.[4] About sixteen hundred years ago, the ancient Polynesians arrived on the islands and subsequently introduced about twenty-nine plant and animal species prior to the arrival of Captain Cook in 1778.[5] During that period of time, the Polynesians had increased the rate of introduction to about one new species every fifty years—a two-hundredfold increase over the natural introduction rate. In modern times, however, about one hundred new species have been introduced to the islands *each year* for the past two hundred years—a millionfold increase in the rate of introductions![6]

In addition to the rate of introductions of species to the islands, humans have also changed the pattern of species introductions. Hawaiian ecosystems developed without large herbivores, carnivores, snakes, and hundreds of plant species that could not make the vast ocean crossing. The arrival of the Polynesians changed this pattern. They introduced plants like taro, breadfruit, and ti and animals like the pig, chicken, and dog. They also accidentally introduced a pest species—the rat. The arrival of Europeans and others—first in large boats and more recently in airplanes—substantially amplified this pattern of introducing species beneficial to humans and accidentally introducing nonbeneficial species.

Modern transportation not only allows the introduction of a far greater number of species but also makes possible the introduction of species from every part of the globe, often with their attendant parasites and diseases. In short, modern human transportation methods and global trade have broken all the natural rules that had allowed for the gradual and selective introduction of species to this isolated island chain. The effect on the Hawaiian Islands has been devastating, as the authors of a recent U.S. Geological Survey report note: "Today, Hawaii is home to approximately 33 percent of the nation's total endangered species, and nearly 75 percent of the recorded extinctions in the United States have involved endemic Hawaiian species. This process continues and may be accelerating."[7]

Unfortunately, Hawaii is but a microcosm of the whole planet. Scientists have pointed out that we are in the process of essentially "homogenizing" the planet to the point that it will become impossible to determine where you are on the globe by simply looking at the species in the area.

Argument: Nature will adapt with time. Even if invasive species do reduce biodiversity, the reduction in number of species will only be temporary. Eventually new diseases, predators, and competition will evolve, and a new diverse ecosystem will develop.

Rebuttal: This statement is technically true. After all, biodiversity rebounded even after the great Permian extinction, when about 95 percent of Earth's species became extinct. However, even after the Cretaceous extinction event some sixty-five million years ago, when about 50–75 percent of the Earth's species, including the dinosaurs, became extinct, it took five to ten million years for biodiversity to reach its former levels.[8] Who would want to ask future generations to wait even hundreds of years for diversification to occur again, much less millions of years? Furthermore, in the current situation, human activity might not allow the Earth's systems to stabilize again unless we find ways to control our population, utilize renewable resources, and stop increasing carbon dioxide and other pollutants into the air and water.

Argument: Invasive species are here to stay, and you can't really control their spread and incredible numbers. Just let nature take its course.

Rebuttal: Of all the skeptics' arguments, this one has the most validity. Undoubtedly, invasive species are here to stay, and their ability to proliferate poses a daunting challenge to even the most dedicated restoration staff and volunteers. Nevertheless, even in a badly invaded region, like Milwaukee County, many natural areas remain that are not beyond recovery. The remaining quality areas need to be vigorously protected and restored as "living natural museums" for future generations. Not only do these remaining quality natural areas provide a glimpse of what our natural areas should be like, they also provide refuge and stopovers for wildlife and a future source of seeds and plant stock of native species. As far as "letting nature take its course," that experiment is already happening in thousands or more acres every day because we, unfortunately, can't protect every pasture, roadside, wetland, and lakeshore from the spread of invasive species. There will always be far more infested areas than resources for invasive species control. Land managers and landowners must carefully prioritize which sites to work on and which species to control.

Argument: Invasive species are too expensive to control.

Rebuttal: The cost of prevention and early control is only a fraction of the cost of control efforts needed once an invasive weed has had the opportunity to go to seed for several years. For example, the Lake County Forest Preserve District in Illinois paid a contractor $150,000 over a five-year period to clear nearly all the common buckthorn (*Rhamnus cathartica*) from a severely infested, 137-acre park.[9] In contrast, it cost The Park People only $500 to clear the same percentage of common buckthorn from a 217-acre park that had significant buckthorn infestation at one edge of the woodland but only occasional specimens elsewhere.

Argument: Changing the vegetation in my local nature preserve will lower my property value.

Rebuttal: Allowing an adjacent preserve or your own property to degrade will probably lower your property value more as the natural beauty and diversity continue to decline. For example, a 1,360-acre ranch in Oregon declined in value from $125–$150 per acre to $22 per acre because leafy spurge had dramatically reduced the productivity of its rangeland.[10]

Argument: I'll lose my visual screen if invasive trees and shrubs are removed.

Rebuttal: The current visual screen can be replaced with native species. Consider replanting the area with native evergreens so that the visual screen is present throughout the year.

Argument: I'll lose my shade if I remove invasive trees from my property.

Rebuttal: This may be true in a prairie or savanna restoration. Consider leaving mature trees along paths or picnic areas.

Argument: I disagree with using herbicides. They can be bad for the environment and toxic to people.

Rebuttal: Consider the analogy of using antibiotics to treat infections or chemotherapy for cancer. Use chemicals only when benefits outweigh risks, and use the safest product in the smallest amount possible. Most people are familiar with the widespread foliar spraying of herbicides in the past in agriculture or along roadsides. In natural areas, herbicide use is targeted to the individual plants or the cut stems of invasive plants.

Argument: Trying to exclude non-native plants and reintroduce native plants in an environment is equivalent to Nazi-inspired gardening.

Rebuttal: This argument is the most bizarre of all arguments made, and yet it received serious press several years ago.[11] The rationale for this argument resides in the fact that the Nazi Party promoted the use of native German plants in the landscaping of the Autobahn and some of the areas they occupied during World War II. Some skeptics like to point out that ecological restoration and natural landscaping specialists emphasize planting native species in nature preserves and home landscapes. The connection between the two uses of native plants is perhaps irresistible but does not follow any rules of logic. The reason restorationists and natural landscapers advocate native plants has nothing to do with an American or racial superiority complex, but is an attempt to maintain or create ecosystems that, in general, promote diversity and are more resilient than landscapes that have been reduced to a handful of ornamental, and often invasive, plants.

NOTES

1. U.S. Office of Technology Assessment. 1993. "Harmful, Non-Indigenous Species in the United States." www.wws.princeton.edu/cgibin/byteserv.prl/~ota/disk1/1993/9325/9325.pdf.

2. Janet Marinelli and John Randall. 1996. *Invasive Plants: Weeds of the Global Garden.* Brooklyn, NY: Brooklyn Botanic Garden. p. 8.

3. D. Pimentel, L. Lach, R. Zuniga, and D. Morrison. 1999. *Environmental and economic costs associated with non-indigenous species in the United States.* Ithaca, NY: Cornell University.

4. Marinelli and Randall. 1996. p. 8.

5. http://gohawaii.about.com/cs/hawaiianhistory1/.

6. Marinelli and Randall. 1996. p. 8.

7. http://www.gap.uidaho.edu/Bulletins/8/bulletin8html/HATIAGA.html.

8. Anonymous. February 1999. "Millennium in Maps: Biodiversity." *National Geographic.*

9. Ken Solis and Lake County (Illinois) Forest Preserve manager. 1998. Personal communication.

10. R. Westbrooks. 1998. "Invasive Plants: Changing the Landscape of America." Washington, DC: Federal Interagency Committee for the Management of Noxious and Exotic Weeds. p. 27.

11. Janet Marinelli. 2000. "Native's Revival: Is Native Plant Gardening Linked to Fascism?" *Plants & Gardens News* 15(2).

Invasive Plant Control Techniques

MOST OF THE INVASIVE plant control techniques described in this book were derived with permission from the *Wisconsin Manual of Control Recommendations for Ecologically Invasive Plants,* published by the Wisconsin Department of Natural Resources Bureau of Endangered Resources in 1993 and again, with updated information, in 1998. Because the information depends on the observations and experiments of land managers working at different sites and in different situations, the level of success of each control technique may vary somewhat across the upper Midwest. Taking this into account, land managers and property owners using this information should evaluate the individual characteristics of their sites to determine which technique will best control the invasive plants there. In many cases, it will be necessary to experiment with different techniques before finding one or a combination of techniques that work. Land managers and property owners should, of course, time their efforts to match the phenological characteristics of the plants they are attempting to control.

Anyone working with invasive species should recognize that while eradication of invasive species may be desirable, it is often not possible. In many situations, a more practical goal is to reduce a weed's population to a level that does not affect the integrity of the native plant community. This will typically allow native vegetation to thrive.

INTEGRATED VEGETATION MANAGEMENT

Integrated Vegetation Management (IVM), which is the use of two or more control methods, is the most effective form of invasive plant control. By combining various preventative and containment techniques, manual or mechanical methods of control, carefully targeted and appropriate herbicide use, biological controls, and the reintroduction of native plant species, land managers and property owners will likely achieve the greatest level of success.

GENERAL GUIDELINES FOR INVASIVE PLANT CONTROL

Following some basic recommendations for invasive plant control will make the effort more efficient and economical, while doing the most good for the environment.

All property owners and managers should:

- Learn to recognize invasive plant.
- Monitor natural areas regularly, especially in the spring, to spot invasions early.

- Maintain a dense cover of desirable native or noninvasive plants to discourage invasive plant seed germination.
- Take steps to control invasive plants as soon as possible after an invasion is discovered. Keep in mind that optimum timing for control will vary from one species to the next.
- Prevent any of the invasive plants in a new infestation from producing seed or otherwise reproducing. Eliminating plants prior to seed development will prevent future generations from developing.
- Minimize soil disturbances when pulling or digging plants. Soil disturbances provide prime areas for invasive plants to grow from seeds that come in from other areas or from seeds in the seedbank.
- Cooperate with land managers and other landowners in the area to make control efforts more successful. Sharing information on the extent of invasions, site characteristics, and various control options is essential.

PRIORITIZING AREAS FOR INVASIVE PLANT CONTROL

To obtain the greatest level of success, it is crucial before beginning any control efforts to develop a plan that establishes priorities, both in terms of species and areas to control. Generally, this involves focusing attention on invasive plants with the greatest potential to damage natural areas or restorations.

- If it is impractical or impossible to remove or treat all invasive plants in a particular natural area, evaluate the site. Areas containing rare and diverse plant species should be noted, with the most valuable areas given top priority. If the natural area is large, it is important to map what kind of invasive plants are present, where they are located, their level of infestation (light, moderate, severe), and the location of important native plants and plant communities. Mapping of plant invasion vectors (trails, roads, waterways) should also be included to understand how plants spread. The map will help prioritize areas for control efforts and can be used in subsequent growing seasons to relocate priority sites, record the level of progress, and make any necessary adjustments to the control strategy.
- Generally, areas in the early stages of invasion (or those on the outlying edges) should be targeted for control before areas of severe infestation. In other words, work from least infested areas to most infested areas. If trying to protect a high-quality area from encroaching weeds, work from the high-quality area outward. (Be careful not to inadvertently carry seeds from infested areas into noninfested areas.)
- A badly infested site with little diversity may be a more important target for invasive plant control if it is an imminent threat to a nearby high-quality natural area. Although a particular invasive plant is located in a badly degraded area, controlling it may be extremely important if it is new to a region and known to have caused major environmental or economic damage in other states or regions at a similar latitude. It may also be prudent to control areas along trails, roads, or streams since hikers, mowers, or flooding can contribute to the spread of invasive plants from these areas.
- Note the location where control efforts are made and return each year to remove additional plants that have emerged in that area. When working on public lands, provide a map or accurate description of the infested site to the site manager.

Controlling Herbaceous Plants

Herbaceous plants are those whose stems die back during the year, typically after seed production or by late fall. Herbaceous plant tissues are usually not woody like those of trees and shrubs, although the roots may continue to produce stems for many years. Herbaceous plants are primarily categorized by their lifespan: annuals live one growing season; biennials germinate the first year, overwinter, produce seed in the second growing season, and die; perennials return every year from rootstocks, producing flowers and seeds most years.

Herbaceous plants can be controlled by a variety of manual, mechanical, chemical, and biological methods. Each method has its pros and cons depending on the plant species and the situation. These methods will be discussed in more detail later in this chapter. There are, however, several general tips for controlling invasive herbaceous plants:

- Preventing seed dispersal is crucial for the control of herbaceous plants. Annuals, biennials, and some perennials will decrease in density over time if they are not permitted to develop and disperse seeds. Sometimes cutting or mowing just before seeds become viable, or collecting seed heads just prior to seed dispersal, will decrease invasive plant density while also preventing soil disturbance. Cutting or mowing does not work with all plants, however, because some will resprout afterward.

- Be aware that most invasive plants can continue to develop viable seed even after the plants have been pulled or cut. Cut or pulled invasive plants that have begun to flower or have flowered should be removed from the area and properly disposed of.

- Dispose of invasive plants properly to avoid transporting their seeds and other propagules to new areas. Composting is not always effective because compost piles may not get hot enough to kill the seeds. Either burn the invasive plants once they have dried (if open burning is allowed in your area) or bury them deep underground in an area that will not be disturbed. If you intend to burn small amounts of collected invasive plants, try filling large paper composting bags, available at hardware stores or garden centers. Leave the tops open (unless seeds are likely to be blown by the wind), and poke holes in the sides to allow air to circulate for faster drying. Protect from rain, then burn after several days. Depending on your state and local laws, sending invasive plants to the landfill with regular garbage may be a disposal option.

- If using herbicides to control herbaceous plants, be sure to use one that will be effective on the specific plant while causing as little environmental damage as possible. Treatment should take place at the most effective time of the year for the target plant (see "Chemical Controls: Herbicides," page 17).

Controlling Woody Plants

Woody plants are typically long-lived trees, shrubs, or vines. Woody plants are categorized as coniferous or deciduous. Conifers or evergreens are cone-bearing, have needle-like leaves, and remain green throughout the winter months. Deciduous woody plants are seed-bearing and have leaves that emerge in the spring, are green during the summer months, and then turn color and drop in autumn. All woody plants have a distinctive barky outer surface. They also add girth to their trunks, branches, and twigs as they grow. While control techniques are often species-specific, there are a few general guidelines for controlling most woody plants.

- Eliminate the large, seed-producing plants first.
- In smaller areas, young trees and shrubs can often be dug or pulled out. Leverage tools for pulling, such as Weed Wrenches, can be helpful in these situations. Larger plants may need to be cut. If the use of herbicides is not possible or desirable, cut trees and shrubs as close to the ground as possible. Resprouts, which can be numerous for some deciduous species, sometimes must be cut repeatedly for several years during the growing season in order to totally kill the plant. Cutting trees and shrubs after full-leaf expansion may result in less sprouting than if they are cut while dormant. If cut late in the summer, some species may resprout but will not be hardy enough to survive a hard frost. Repeated cutting, without the help of herbicides, is often impractical for large populations.
- Most deciduous woody plants will resprout after being cut unless an herbicide is applied to cut surfaces. Herbicides do not need to be applied to conifers because they will not resprout after being cut.
- Clonal species, such as willows and aspen, will send up new shoots after being cut, sometimes several yards away from the cut stems.
- Several techniques are available for cut-surface herbicide applications, but their use depends on the species, time of the year, and site situations.
- After cutting, it is usually best to leave cut trees and shrubs on site since physically removing them may be difficult. Create small brush piles with cut trees and shrubs to provide shelter for wildlife or for pickup and disposal elsewhere. Larger piles of cut material can be created and burned (see "Burning Brush Piles," page 16). Brush removal is acceptable if suitable machinery and a disposal site are available.
- Girdling—removing a two- to five-inch ring of bark from a tree or branch—can be used to eliminate some woody plants. This technique is easiest in late spring when the sap is flowing and the bark readily pulls away from the sapwood. Trees die slowly over the course of one or two years. If girdled too deeply, trees will respond as if they had been cut and resprout from the base or roots. Girdling is not effective on all species and may lead to sprouting or suckering; check with experts before using this technique. See "Girdling Tools" at the end of this chapter.

- Use safety precautions. Wear gloves, eye protection, and protective headgear in brushy areas. Hire properly trained and insured individuals for cutting large trees and shrubs. When working with chainsaws or brush saws, wear protective chaps, always work with another person, and follow proper operating procedures. Know and understand how to properly and safely use herbicides.
- Cleared sites require periodic monitoring to prevent the reestablishment of weedy plants. Be prepared to manage the rapid growth of seedlings after unwanted trees and shrubs have been removed and more sunlight begins to reach the forest floor.

GENERAL TECHNIQUES

There are a wide variety of basic techniques for controlling invasive plants. These techniques include manual controls, such as hand-pulling, digging, flooding, mulching, and burning; mechanical controls, such as pulling, cutting, girdling, tilling, mowing, and chopping with various tools and machines; chemical controls; and biological controls. The following section discusses burning, herbicides, and biological controls.

Use of Fire to Control Invasive Plants

If using fire as a control technique, know and follow local burning regulations. Grass or prairie fires can spread rapidly. Consult with natural areas management experts at nature centers or with The Nature Conservancy for training opportunities before attempting this method of control. Burn permits are required in many areas. Contact your local fire department and natural resources office for more information.

Controlled Burns

Controlled or "prescribed" burns are used to reduce invasive and woody plant density and competition, stimulate the growth of native plants, return nutrients to the soil, promote germination of dormant seeds, and enhance wildlife habitat. These burns are called "controlled" or "prescribed" because they are done only under specific weather- and fuel-related conditions that ensure an effective burn and the safety of the burn crew and the surrounding area. Purposely set in plant communities that have evolved with fire, such as oak woodlands, prairies, savannas, and sedge meadows, controlled burns can kill or set back certain invasive species that do not tolerate fire. Burns are usually conducted in midspring or fall. If early blooming wildflowers are present, it is often best to burn in very early spring or late fall to avoid damaging them.

In natural areas that are badly infested with invasive plants, controlled burns may initially need to be done for several years in a row to reduce the weed seedbank and stimulate native species. Burning this frequently is not generally recommended in healthy native plant communities because important insect pupae and eggs may also be destroyed. Burning one-third to one-half of a natural area each year on a rotating basis is usually the preferred management strategy and will lead to increased plant and insect diversity.

Conducting controlled burns should not be taken lightly. Burning is a dangerous activity that requires planning, coordination, equipment, and trained personnel. It also requires an understanding of how fuel conditions and weather conditions, such as humidity, temperature, wind direction, and wind speed, affect a burn.

Anyone considering using burning to control invasive plants should determine if fire is the best management strategy, in terms of its effectiveness in controlling a given species and in terms of the risks and costs involved. If fire is chosen as a management tool, a site fire-management plan should be developed and include the following: site information, fire management justification, fire management goals, site-specific fire operations, smoke-management plan, and an assessment of how adjacent neighbors and the community will react to the use of fire. A prescribed burn plan should then be written for each burn unit. This plan should specify the weather conditions, fuel moisture level, personnel and equipment requirements, and people/agency contacts that must be met before a prescribed burn can take place. The plan should also include information about what to do if a fire escapes containment, including emergency phone numbers.

The proper equipment and clothing is very important for conducting safe prescribed burns. The following are necessary: walkie-talkies/radios, cell phones, drip torches, backpack pump sprayers, containers of water, Nomex clothing, hard hats and face shields (especially when working in wooded areas), cloth masks or respirators for preventing smoke inhalation, leather gloves, leather work shoes, and fire rakes and/or leaf blowers for creating fire breaks. Foam units are expensive but effective for quickly putting out smoldering tree trunks and stumps when burning wooded areas. Many larger land management agencies also have pumper trucks or ATVs that are equipped with water tank and hose units.

The following resources contain more information about using controlled burns for invasive weed control:

> *How to Manage Small Prairie Fires* by Wayne R. Pauly. To order, contact Friends of Dane County Parks; also available from several prairie nurseries in the region.

> "Conducting Burns" by Wayne R. Pauly in *The Tallgrass Restoration Handbook*, Society for Ecological Restoration, edited by Stephen Packard and Cornelia Mutel, 1997, Island Press, Covelo, CA, www.islandpress.org.

> The Nature Conservancy's Web site: tncweeds.ucdavis.edu/handbook.html.

Spot Treatment with Fire

Sometimes large-scale burns are either inappropriate or cannot be done because of high fuel moisture or a lack of fuel to carry a fire. In such situations, spot-treating plants that are vulnerable to fire may be preferred. A propane torch with a long wand works well in these situations and can be used to treat

Every year, the costs associated with non-native weeds approach and exceed $35 billion in the United States alone. (D. Pimentel, L. Lach, R. Zuniga, and D. Morrison, "Environmental and Economic Costs Associated with Non-indigenous Species in the United States," Cornell University, Ithaca, NY, 1999)

individual plants or small groups of plants. This technique works especially well on seedlings and young saplings, and is generally less labor-intensive than hand-pulling and less expensive than herbicide treatments. Spot burning often only top-kills more mature plants.

Propane torches suitable for spot-burning invasive plants are available in many garden centers and can be ordered online. Several propane torch devices are available and range in size from small handheld units to larger units requiring a five-gallon propane tank.

To burn invasive plants with a propane torch, the operator passes the flame over the plant for less than one second, which is enough time to boil the water in the plant. Boiling the water within the plant will cause it to droop immediately and die within a couple hours.

Some municipalities will not allow other types of prescribed burning but will approve burn permits for spot-treating plants with a propane torch. Be aware, however, that propane torches are dangerous to use, and there are several safety hazards that should be observed.

- Wear fire safety gear (Nomex fire suit, hard hat with face shield and Nomex ear/neck protector, leather gloves, and leather boots).
- Make sure that the pressure relief valve on top is pointing *away* from the person carrying the pack. This valve is designed to release pressure by venting gas. If the tank overheats, the vented gas can ignite and will burn the applicator if the valve is pointing inward.
- Always be cautious where you set the torch down because the tip of the torch gets very hot. Cool it down with water, if possible.
- Do not use propane torches on dry vegetation or when windy conditions exist unless you have made adequate preparations to contain a runaway fire.

For more information, see tncweeds.ucdavis.edu/tools/rdtorch.html and doityourself.com/gardentools/weedcontrolflamers.htm.

Burning Brush Piles

Sometimes, weed-infested natural areas may lack enough dried grass, leaves, and other vegetation to support a fire hot enough to eliminate or set back encroaching trees and shrubs. If this is the case, woody species may need to be cut. Cutting in winter, when there is a good layer of snow and the ground is frozen, will help protect the underground root systems and groundlayer plants from trampling and dragging. Snow cover will also help limit the impact on the soil and soil microorganisms from the intense heat generated when burning the piles of brush. Soil and soil organisms should be rescued prior to setting fire to brush piles that are directly on top of the soil. They should be replaced once the burn pile is consumed in order to replenish the soil with its original microorganisms.

Well-managed land is the best defense against the spread of weeds. (Jerry Asher, "War on Weeds: Winning It for Wildlife," Bureau of Land Management, 2000)

Chemical Control: Herbicides

Author's note: Although herbicide trade or brand names are mentioned in this book, they are made to assist the reader in their research and work and do not represent an endorsement of a particular herbicide or control method. Other products with the same or equivalent ingredients may be available. Exclusion of an herbicide does not imply that its use is unacceptable. The information provided below is based on information available at the time this book was written; herbicide labels change over time. Precautions discussed may be incomplete and do not supersede label directions. Herbicide users must read and follow directions on the product label.

The Proper Use of Herbicides

In a perfect world, chemicals would never be needed to control invasive plants. But this is not a perfect world, and the risks of using herbicides must be weighed against the mounting negative effects that invasive plants are having on natural areas and other areas of importance. The decision to use an herbicide is much like deciding whether to use an antibiotic against a virus. The risks of side effects from the antibiotic are outweighed by the benefits of saving a patient from infection—especially if the safest and most appropriate antibiotic is used.

For all practical purposes, some invasive plants cannot be controlled without the use of herbicides. Buckthorn, for example, is often cut if it is too large to pull. If the stump is not treated with an herbicide, it will vigorously resprout and become a more multistemmed tree or shrub than before. Other invasive weeds, such as Canada thistle, have root systems that are too extensive to dig or pull out of the ground. (Digging or pulling will also leave an area of exposed soil that will offer a prime site for reinfestation because a weed seedbank is likely present.) In other situations, the sheer number of invasive plants may require the use of herbicides because manual and mechanical controls would be overly expensive in terms of the time, cost, and/or labor involved.

ALWAYS READ THE ENTIRE LABEL before purchasing and using an herbicide. The label is a legal document that must be read and obeyed. Moreover, it provides important information on the proper use, rate and timing of application, transportation, storage, cleanup, and emergency procedures for the particular herbicide being used. *Label directions supersede the guidelines listed in this book if there is a conflict in information.*

- Use herbicides only when necessary. Every control method has its advantages and disadvantages. In general, other weed controls should be considered first, whenever it is safe and feasible to do so. Herbicides should be considered for use on species or plant populations that are too difficult or expensive to control otherwise, or when manual or mechanical control would be unsafe or likely cause more environmental damage than responsible chemical control.

- Be sure you are adequately trained to use herbicides. Contact your state's agriculture department or local extension office for more information about herbicide applicator training classes and restrictions for herbicide use. Restrictions vary from state to state. Although few herbicides require certification for noncropland use, check the product label to determine whether certification is required. Certification may also be required if someone is applying any herbicide on a for-hire or contract basis.
- Use herbicides at the right time of the year to control invasive plants and avoid harm to desirable species. For example, some invasive plants can be treated in late fall, winter, or early spring when most native vegetation is dormant and less likely to be damaged by herbicides.
- Place commercially available signs around the perimeter of a site to notify people that the area has been treated with herbicide. The signs should be removed when the threat of exposure to the herbicide has passed and it is safe to reenter the area.
- Avoid herbicide contact with eyes, skin, or clothing. Wear protective glasses and chemical resistant gloves when handling all products, and wash hands frequently. Always wash before eating, smoking, or using the restroom. Avoid touching bare skin with contaminated equipment or clothing. Do not inhale or swallow herbicides. Carry water and eyewash with you in the field in case of an accident.
- Contact the manufacturer and request a Material Safety Data Sheet (MSDS) for the specific herbicide being used. The MSDS will provide you and health care professionals with the appropriate emergency and medical measures to take in the event of accidental exposure. The Internet also has printable product labels and MSDS information (see "Herbicide Information," page 27).
- Watch for and understand "signal" words on labels that indicate an herbicide's relative toxicity (according to the type of exposure—oral, inhalation, or skin): *Danger* (highly toxic), *Warning* (moderately toxic), and *Caution* (slightly toxic).

Transportation and Storage of Herbicides

- Herbicides should be stored, preferably in a locked cabinet, in their original container in good condition and with an attached legible label.
- Do not store herbicides with food, animal feed, or personal protective equipment.
- Storage of mixed solutions can reduce the effectiveness of some herbicides and create disposal problems. If it is necessary to store herbicides after mixing, they should have the exact herbicide label(s) on them. There should also be a sign indicating that they contain mixed solutions, and the date of the initial mixture. Mixed chemicals should be used within a few days.
- Transport secured herbicide containers outside the passenger compartment of a vehicle.
- Always carry spill response equipment in the event of an accidental herbicide spill.

Mixing and Loading Herbicides

- READ THE ENTIRE LABEL before using any herbicide.
- When mixing or loading herbicides, wear protective clothing to reduce the risk of exposure. Read the label for the personal protective equipment required.
- Keep drinking water, decontamination water, rinse water, and mixing water in separate containers.
- Mix and load on an impermeable surface, such as concrete or blacktop, or by placing the application equipment into a chemical-resistant tub or basin for containment. Do not mix within one hundred feet of surface water, a well, or storm drain. Should a spill occur, recover it immediately and report it to the appropriate spill response agency. Keep the agency phone number along with first aid and emergency guidelines at your mixing site.
- Avoid mixing more herbicide than is needed.
- Appropriate diluents (water or another liquid) should be added when it is necessary to weaken an herbicide and achieve the recommended concentration for an application method or for use on a particular weed. Be aware that an already diluted herbicide cannot be made more concentrated. For example, Roundup, sometimes sold as a 1.92% a.i. (active ingredient) glyphosate solution, is often used to kill lawn grass or garden weeds. A more concentrated formulation of glyphosate, such as Roundup Pro, must be used to achieve the 20% a.i. concentration needed for cut-stump treatment of most woody plants. Be sure to read the label carefully in order to calculate the correct amount of diluent needed to achieve a desired concentration.
- By law, herbicide applications must be consistent with label directions. In some instances, this book may suggest a concentration lower than that recommended on the product label based on reported research. It is acceptable to use a weaker solution than that recommended. However, this does not guarantee that the lower concentration will be effective in every instance. In addition, the manufacturer may not stand behind its product if this is done.
- Do not mix one herbicide with another unless the combination is listed on the label.
- Colorants or marker dyes can be added to some herbicides to help the applicator see what areas have already been treated. Besides helping the applicator avoid chemical contact, colorants can help save time and money by avoiding retreatment of plants and overuse of herbicides. Follow label instructions.

Herbicide Applications

- Wear appropriate personal protection equipment as per label or guidelines listed above.
- Do not overapply herbicides so runoff from the plant occurs or contact is made with nontarget species.
- Work away from areas already treated.
- Do not apply herbicides beyond the boundaries of the target area.
- Prevent herbicide drift (see "Particle Drift," page 22).

Methods of Herbicide Treatment

Although there are other ways to apply herbicides, the six commonly used methods for invasive plants control (cut-stump, basal bark, frill, foliar, injection, and automatic application treatments) are described here.

Cut-Stump Treatment Restorationists use the cut-stump technique to eliminate most woody plants. It involves applying an herbicide, such as glyphosate (e.g., Roundup Pro) or triclopyr (e.g., Garlon 4 with penetrating oil or Pathfinder), to freshly cut stumps or stems using a low-pressure handheld sprayer, a sponge paint brush, or a contact lens solution bottle (only for glyphosate applications). Adding an appropriate dye to the solution can help the applicator see which stumps have already been treated. To eliminate a colony of woody or herbaceous species that reproduces by underground runners, every stem in the colony (or clone) must be cut and treated to discourage resprouting. Use glyphosate approved by the Environmental Protection Agency for aquatic applications, (e.g., Rodeo, Aquamaster, Glypro) when applying herbicide to cut fen buckthorns in standing water. A permit from your state's natural resources department may be required to apply herbicides near open water.

A crew of two or three people works best when using the cut-stump treatment on large populations of invasive woody plants. One person cuts the unwanted plants, the second person applies the herbicide, the third person hauls the brush out of the way.

Application: Apply the herbicide to the entire cambium ring of the cut stump of woody species. The rest of the stump does not need to be treated because the inner heartwood will not conduct the herbicide to other areas of the plant. It is difficult to avoid this inner area when treating small trees or stems. When using triclopyr, the herbicide should cover the sides of the cut stump down to the root crown.

Time of Treatment: Since glyphosate is not mixed with a wood-penetrating oil, fall is usually the best time for cut-stump treatments of glyphosate; at that time of the year the plant's sap is flowing toward the roots, which allows for maximum herbicide absorption. Spring (mid-March to mid-May) is the least effective time for cut-stump treatment with glyphosate because the sap is flowing away from the roots. Stumps should

be treated as soon as possible and not more than thirty minutes after cutting.

Cut-stump treatments with triclopyr can take place any time of year, although spring treatments are the least effective. Herbicide application to live stumps can be delayed after cutting, but this can make the stumps harder to find.

Temperatures: Glyphosate may not be effective in cold temperatures when the sap is not running. Solutions and spray nozzles may also freeze. Triclopyr in an oil carrier can be used during extreme cold but should not be used during hot spells, especially at temperatures of 80°F or warmer. At high temperatures, solutions may become volatile putting the applicator and nontarget species at risk. Follow label directions to avoid injurious spray drift.

Mixing Rates for Cut-Stump Treatment: A 20–25% a.i.(active ingredient) solution of glyphosate will control most woody species when used as a cut-stump treatment. If using a glyphosate product that has 41% a.i., mix one part glyphosate with one part water to obtain a 20% a.i. solution.

A 12.5% a.i. solution of Garlon 4 (triclopyr) diluted with penetrating oil has generally been effective as a cut-stump treatment for controlling woody species. Garlon 4 has a 61.6% a.i. concentration. Mix one part Garlon 4 with four parts penetrating or bark oil to obtain a 12.5% active ingredient solution.

Pathfinder II, another triclopyr herbicide, comes premixed and ready for use.

Basal Bark Treatment Basal bark treatment refers to applying herbicide (usually Garlon 4 mixed with penetrating oil or Pathfinder II) in a six- to fifteen-inch band around the entire trunk of a tree or the stems at the base of a shrub. The root collar around the base of the plant may be treated as well. If the solution is not premixed, a 12.5% a.i. solution has generally been effective. A weaker solution may work on smaller trees and shrubs, or stems less than two inches in diameter. This method of treatment is inconsistent on stems or trees larger than six inches in diameter at the base and should not be used if snow prevents coverage to the ground line. Applications of ester formulations should be avoided when temperatures are hot, especially when greater than 80°F. Read the label to avoid conditions where injurious spray drift may result.

Table 2.1 **Cut-stump Treatment: Comparing Glyphosate (Roundup Pro) and Triclopyr (Garlon 4)**

Frequently Asked Question	Roundup Pro	Garlon 4
What is it mixed with?	Water	Penetrating bark oil*
Where is it applied?	Cambium layer just inside the bark on a cut stump.	Cambium layer and down sides of cut stump to root crown.
What time of year should it be applied?	Fall is the most effective time to apply. Avoid application during spring sap flow or cold winter months.	Can be applied anytime; most effective in fall, may be less effective in spring.
What are the temperature restrictions for application?	May not be effective during cold winter months when temperatures are below freezing.	May vaporize and drift when temperature is greater than 80°F.
How quickly after cutting must herbicide be applied?	As soon as possible or within thirty minutes.	Anytime, but shortly after cutting is recommended.
What plants will be affected?	Is nonselective, may kill or harm any plant it touches.	Broadleaf specific, should not harm grasses if used according to label directions.

Note:

*Although the product label may list diesel fuel, No. 1 or No. 2 fuel oil, or kerosene as possible ingredients to mix with Garlon 4, penetrating oil or bark oil is generally considered less harmful to the environment.

With basal bark treatment, numerous trees or shrubs can be treated in a short period of time without cutting. This may be important when time and personnel are limited, large areas of land are involved, and infestations are severe. Treated trees and shrubs generally fall down over time, or they can be cut down once they are dead.

Foliar Treatment Foliar herbicide application refers to applying an herbicide to the green leaves of an invasive plant. This is usually done with a sprayer unit or some type of wick applicator. A wide variety of sprayers exist including small handheld sprayers, backpack sprayers, spray units mounted on ATVs or in truck beds, and sprayer units pulled behind tractors. When applying a foliar spray, the herbicide should thoroughly wet fully opened or expanded leaves but should not be so heavy that it drips off. Spot spray whenever possible rather than treating uninfected areas.

Wick application involves applying herbicide to foliage with an absorbent material, like a sponge, that has been secured to a long handle. This method avoids spray drift and limits exposure to nontarget plants. Cotton swabs dipped in herbicide can be used for applications to very small areas.

When leaves are fuzzy or waxy, some herbicides can be mixed with a surfactant to enable them to be more easily absorbed.

Some non-native weeds remain green during the winter months, begin to leaf out earlier in the spring, or hold their leaves later in the fall than most native species. Such species can be treated with a foliar spray at these times as long as temperatures are 50°F or above. This method not only provides effective control, it eliminates the possibility of herbicide contact with dormant native species. If you plan to spray in the spring while native plants are still dormant, it may be helpful to burn or mow the area in the fall before treatment to remove the heavy layer of dried vegetation and create more exposure for new invasive plant growth.

Foliar treatments should not be made in windy conditions or when wind direction will put people, animals, or desirable plants at risk of exposure from herbicide drift. Foliar herbicide applications are typically not effective during periods of drought or once the leaves turn color and stop photosynthesizing in the fall.

Girdling and Frilling Treatments Girdling involves cutting or otherwise removing a band (one to two inches wide on small-diameter trees, six to eight inches wide on large-diameter trees) of bark around the entire trunk of the tree to interrupt the flow of sap between the roots and the crown of the tree. Girdling can be done with an ax, hatchet, chainsaw, or other tools specifically designed for the task. If done without the use of herbicides, girdling is most effective if done prior to spring sap flow.

Frilling is a variation of girdling that involves cutting a continuous ring of overlapping notches through the bark around a tree trunk with an ax, hatchet, or chainsaw, preferably within twelve inches of the base.

While both girdling and frilling are relatively effective at killing trees, the effectiveness and speed of the control is increased when the exposed cambium is saturated with herbicide. The same herbicides and concentrations used for cut-stump treatments will work on girdled and frilled woody plants.

Injection Treatment Injection treatment is a fairly new method of herbicide application that uses a lance to inject a .22-caliber cartridge filled with a gelatinous, water-soluble form of herbicide into the trunks of trees and larger shrubs. The number of cartridges per plant depends on the diameter of each trunk. Some land managers report less success when using the recommended number of cartridges on multiple-trunk trees.

Although the injection lance and cartridges are costly, injection treatments require considerably less labor than other treatment methods because a large number of plants can be treated fairly quickly. In addition, the injection system is safer for the operator to use than other types of chemical treatments, and there is virtually no chance of harming other plants. There are several types of herbicides available in cartridges, including glyphosate, imazapyr, picloram, hexazinone, and triclopyr. According to the manufacturer's information, each herbicide will treat a specific list of trees and shrubs.

Automatic Application Automatic herbicide application involves the use of machinery that will cut woody plants and then treat them with herbicide. This equipment is currently available only to a limited extent. Plans are underway to perfect and market chain and brush saws that apply herbicide to the cut surfaces as they cut. One such tool is the Brown Brush Mower. Pruning shears that do this are now available along with a brush hog tractor that can simultaneously cut woody stems while applying an herbicide to them. For more information, see "Weeding Tools" section, page 30.

Preventing Herbicide Drift

Herbicide drift, one of the most common problems encountered when applying herbicides, is the movement of herbicide to areas beyond the one targeted for treatment. Besides killing or severely damaging desirable plants, herbicide drift can harm people (including the applicator), domestic animals, wildlife, and insects. Soil and water can also be contaminated. Applicators should know what kind of plants and/or land uses are adjacent to the treatment area. They should also be monitoring weather conditions, especially for wind speed, wind direction, and ambient temperature.

Drift can occur in the form of spray particles or as vapors.

Particle Drift Particle drift occurs when wind or air currents move spray particles (or droplets) that contain herbicide beyond the intended control area. To prevent particle drift:

- Do not spray herbicides in windy conditions. Wind speeds are usually lower early in the morning or in the evening.
- Be aware of wind direction and the potential impact that herbicides may have on plants and animals downwind. Use of a handheld wind gauge may be helpful.
- Keep spray nozzles or booms as low as possible (or as low as recommended by the manufacturer) to minimize wind effects.
- Increase spray droplet size. As the size and weight of spray particles increase, the potential for particle drift decreases. This can be done by:

  - Decreasing spray pressure and using high-flow rate nozzles.

  - Using water diluents when appropriate. This results in larger particles and less drift than oil diluents.

  - Adding thickening agents to herbicide solutions to help produce larger droplets. Check with an herbicide dealer for more information and to see if thickening agents can be used in a particular situation.

- Avoid applying herbicide during periods of temperature inversion. Temperature inversions are characterized by cooler temperatures at the soil surface and warmer temperatures above. Spray droplets become trapped below the warmer air and begin to move laterally, potentially damaging plants outside the target area. Inversions typically dissipate as the soil warms during the day, at which time drift becomes less of a problem.

If small areas of herbaceous plants are to be spot-treated with an herbicide:

- Place a large plastic soda bottle with the bottom cut off over small, individual weeds and then spray the plant through the top of the bottle. This will help isolate the spray from surrounding plants. Using a piece of cardboard, or something similar, as a shield will also help protect nontarget plants from particle drift.
- Cut the target plant back somewhat before spraying while retaining many of its leaves will make herbicide contact with nontarget plants less likely.
- Weeds growing near desirable plants can sometimes be cut at their base followed by the application of a strong solution of herbicide (e.g., Brush-B-Gon, Roundup Pro) to the cut surface. Check the herbicide label to see if the formulation is appropriate for this technique.

Vapor Drift   Drift can also occur when volatile herbicides go from a solid or liquid state to gas (vaporize or form fumes). Vapors can escape farther from the target area and drift for a longer time than spray droplets. Garlon 4, Crossbow, and 2,4-D ester are among the herbicides that may vaporize during or after application. Dicamba is also prone to vapor drift but less so than the previously mentioned herbicides.

To reduce the potential for vapor drift:

- Spray when the air is slightly unstable and there is a light, but steady, breeze. Avoid spraying in very calm conditions because it is difficult to know what direction the herbicide will drift when air movement finally begins.
- Spray when ground air temperatures are less than 80°F because air tends to rise rapidly at higher temperatures, carrying herbicide vapors across long distances. Herbicides are also more likely to volatize at higher temperatures. During the summer months, spraying in the late morning or early evening is recommended.
- Ester formulations should be used only with extreme caution in warm summer months. They are best for

late fall, winter (cut-stump or basal bark treatments), or early spring applications when temperatures are cool. Used at these times, esters are more effective than amine formulations (which are nonvolatile) because they are more readily absorbed by plant tissues. Check the herbicide label under "active ingredients" to see if the herbicide is an ester formulation.

- Be aware that spraying during periods of low humidity and high temperatures can lead to evaporation, reduce the droplet size of herbicide sprays (primarily fine sprays), and result in drift.

Herbicide Equipment Cleanup

- All empty herbicide containers should be triple-rinsed, using 10 to 20 percent of the container volume for each rinse. Empty the container completely. Save and use the rinsate from cleaning herbicide containers and measuring equipment in future mixes. Be careful, however, not to make the herbicide solution stronger than the label allows. (Tip: Add only half the water needed to an herbicide for an application, add the rinsate from cleaning measuring equipment and containers, then fill with the remaining water needed.)
- At the end of the day, all personal protection equipment should be washed with mild soap and water. This water can be sprayed through equipment as a rinse treatment. If concentrate is spilled on clothing, discard the items. Wash herbicide application clothing separately from other laundry. Do not wear application clothing again without washing it first.
- Shower and change into clean clothing as soon as possible following herbicide applications.
- Check with your local waste disposal contractor or state natural resources department for disposal recommendations for herbicide containers.

Herbicide Terminology

Herbicides have three names: trade name, common name, and chemical name (formula). For example:

Roundup® (trade name)
Glyphosate (common name)
N-(phosphonomethyl) glycine (chemical name)

The following are some commonly used terms to describe herbicide or herbicide-related products:

Adjuvant: A chemical added to make the herbicide more effective or safer.

Broadleaf specific: Indicates that broadleaf plants are affected by the herbicide. Monocots, such as grasses, sedges, cattails, rushes, or reeds, are typically not harmed if a broadleaf-specific herbicide is used according to label directions.

Contact: Refers to a type of herbicide that kills only the parts of the plant where direct contact is made.

Nonselective: Indicates that the chemical may kill or seriously damage any plant it comes in contact with.

Pre-emergent: Refers to a type of herbicide that is applied to the soil surface and kills germinating seedlings.

Selective: Indicates that the chemical is only effective against a certain range of plants. Other species should not be seriously injured by the herbicide if contact is made.

Soil residual activity: Indicates that an herbicide will continue working in the soil for a period after the initial application.

Surfactant: An adjuvant added to herbicides to reduce the surface tension of water and enhance the ability of the herbicide to be absorbed by the treated plant.

Systemic: Indicates that the chemical will affect all parts of the plant, including roots and rhizomes, after it is absorbed by the foliage or cambium.

Use this herbicide information to your advantage when possible. For example, using a broadleaf-specific herbicide, such as triclopyr or 2,4-D, to control broadleaf weeds in a prairie will allow grasses to survive and continue competing against the invasive plants. Herbicides that kill only grasses will allow broadleaf plants and sedges to survive. If a site is severely degraded with few desirable plants, a nonselective herbicide that can potentially kill all plants in the area may be acceptable to use.

Herbicides Used for Invasive Plant Control

The following is a list of the common names of most herbicides discussed in this book, examples of trade names of products containing these active ingredients (other products with similar formulations may be available), and herbicide characteristics. Some herbicides listed are available only in two-and-a-half-gallon

containers, making them very expensive or impractical for the small property owner or land manager. In such cases, use of an alternative herbicide may be necessary. The comments given with each herbicide apply when the herbicide is used as directed on the label. (See "Methods of Herbicide Treatment," page 20, for details about various kinds of treatment.)

Clopyralid

Trade names: Transline (for noncrop areas), Stinger (for crops)

Comments: Systemic, water soluble, highly selective foliar spray. Effective against species in the Buckwheat (knotweed), Legume (pea), Violet, Nightshade (tomato), and Composite (sunflowers, asters, goldenrods, coneflowers, and other daisylike flowers) families. Safe for use near conifers and most trees and shrubs. May be used on grazing land; has little effect on grasses and other monocots. Do not apply in areas where soils have rapid permeability (such as loamy sand to sand, or soils lacking clay particles or organic matter) or where the water table is shallow since groundwater contamination may occur. Practically nontoxic to slightly toxic to animals. Has soil residual activity.

Dicamba

Trade names: Vanquish, Banvel, Banvel CST, Veteran CST, BK 800 Trimec (contains 2,4-D)

Comments: Systemic, water soluble, broadleaf specific. Will kill broadleaf plants before and after sprouting. Can be applied to leaves or the soil. CST brands can be used for cut-stump treatment. Effective against legumes. Avoid injurious spray drift. More soil residual than 2,4-D and clopyralid. Does not bind tightly to soil particles and is highly mobile. Do not apply where groundwater depth is shallow or to sandy soils with less than 3 percent organic matter since groundwater contamination can occur. Poses little threat to wildlife.

Fluridone

Trade names: Sonar A.S., Sonar SRP

Comments: A selective, systemic aquatic herbicide for management of aquatic vegetation in fresh water ponds, lakes, reservoirs, drainage canals, and irrigation canals. Absorbed from water by plant shoots and from the soil by roots of aquatic vascular plants. Exposes the plant's chlorophyll to ultraviolet light, which causes the leaves to turn white. Rapid water movement or any condition that results in rapid dilution of this product in treated water will reduce its effectiveness. Consult with appropriate state or local water authorities before using. Use permits may be required.

Fluazifop-p-butyl

Trade names: Fusilade II, Ornamec
Comments: A selective, systemic, postemergent herbicide for control of most grasses. Should not harm broadleaf species. Often used with an oil adjuvant or surfactant to increase efficiency. Low toxicity to birds and mammals but can be highly toxic to fish. Do not use in areas of standing water. Binds easily with soil particles, limiting leaching and runoff. Often best to apply when native grasses are dormant and non-native grasses are vulnerable.

Fosamine

Trade name: Krenite S
Comments: A selective, nonvolatile, contact, water-soluble bud inhibitor. Used as a foliar spray or cut-stump treatment to control and/or suppress growth of woody plants and brush. Plants sprayed in summer or fall will fail to leaf out the following year and then die because they are unable to photosynthesize. Thorough coverage is essential. Effective on a limited number of herbs, including field bindweed and leafy spurge. Does not affect grasses, sedges, and most herbaceous plants. Almost no soil activity. Has a very low level of toxicity to people and wildlife. Rapidly degraded by soil microbes. Do not use on food or feed crops. Use may be restricted in some states.

Glufosinate-ammonium

Trade name: Finale
Comments: Nonselective, contact herbicide. Typically sprayed on leaves and green stems. Best used on annuals and biennials. Degrades rapidly. Low toxicity. No soil residual activity. No uptake occurs through roots or mature woody bark.

Glyphosate (for use on dry sites)

Trade names: Roundup, Kleen-up, Roundup Pro, Touchdown, Mirage. Glyfomax, ClearOut, Razor Pro
Comments: Nonselective, systemic, water-soluble, widely used and widely available herbicide. Can be used as a foliar spray on herbaceous or woody plants. Some products can be used for cut-stump or cut-stem treatments. Strongly adsorbed by soil particles preventing excessive leaching and root uptake by nontarget plants. Little or no soil activity. Although glyphosate by itself has relatively low toxicity for fish, birds, and mammals, the surfactants in some formulations are highly toxic to aquatic organisms. Typically sold in two-and-a-half-gallon containers, although smaller containers of specially formulated, ready-to-use products (Roundup Super Concentrate) are readily available.

Glyphosate (for use in areas of standing water)

Trade names: Aquamaster, Accord, Rodeo, Aqua Neat
Comments: Nonselective, systemic, water-soluble herbicide formulated for use on above surface plants in aquatic areas. Rapidly dissipates in water by adsorption to suspended and bottom sediments. Can be used as a foliar spray (a surfactant, such as L1700, may be needed for consistent performance) or for cut-stump (e.g., glossy buckthorn) or cut-stem treatment in areas of standing water if label suggests. Minimal soil residual activity. Consult with appropriate state or local authorities before using in areas of standing water. Use permits may be required.

Imazapic

Trade name: Plateau
Comments: Selective, systemic herbicide. Selectively controls some broadleaf weeds and certain annual and perennial grasses. Warm-season prairie grasses, such as big and little bluestem, sideoats and blue grama, buffalograss and Indiangrass, tolerate its use. Certain kinds of cool-season grasses, wildflowers, and legumes also tolerate low application rates. Should not be used on food or feed crops. Do not use in areas where soils are permeable, particularly where the water table is shallow, since groundwater contamination can occur. Methylated seed oil or a vegetable oil concentrate may need to be added to make it more effective. Relatively nontoxic to mammals, birds, and amphibians. Persists in soil. Controls leafy spurge and reed canary grass.

Imazapyr

Trade names: Arsenal, Chopper, Contain, Stalker, and for aquatics, Habitat
Comments: A nonselective, broad-spectrum herbicide that disrupts a plant's protein synthesis and interferes with cell growth and DNA synthesis. Plants tend to die slowly (several weeks). Controls annual and perennial

grasses and broadleaved plants, brush, vines, and many deciduous trees. Can remain active in the soil for six months to two years. Has little potential of leaching into groundwater. Can be applied as a pre-emergent, but is most effective as a post-emergent herbicide. Low toxicity to animals, including fish and insects. Do not over-apply since excess herbicide can reside in soil for extended time periods, killing no-target species. Can cause severe, irreversible eye damage.

Metsulfuron-methyl

Trade name: Escort

Comments: A dry, selective, systemic herbicide that is mixed with water and a surfactant. Should not harm grasses. Effective on many herbaceous weeds, including legumes and woody plants in noncrop areas and conifer plantations. Has soil residual activity. Moderate eye irritant. Very low or low toxicity to mammals and other wildlife.

Pelargonic acid

Trade name: Scythe

Comments: Fatty-acid-based, nonselective, contact, water-soluble herbicide that is applied to foliage. Suppresses annual weeds and grasses by destroying green foliage and top-kills perennial species. Most effective on weeds less than six inches in height. Depending on plant size and species, regrowth may occur and require additional treatment. Rapidly degrades. Provides no residual weed control.

Picloram

Trade names: Pathway, Tordon RTU (both products contain 2,4-D)

Comments: Systemic, water-soluble, ready-to-use herbicide. Effective in cut-surface applications for killing unwanted trees (e.g., black locust) and preventing undesirable sprouting of cut trees in forests and noncropland areas, such as fence rows, roadsides, and rights-of-way. (Picloram is not suggested for other forms of weed control.) Has a tendency to drift. Do not apply to soils that have a rapid permeability (loamy sand to sand) or where water tables are shallow since groundwater contamination can occur. Can move with surface runoff water and remain toxic to plants for a year or more if it gets into the soil. Do not apply to lawns or near desirable trees, shrubs, and broadleaf plants.

Sethoxydim

Trade names: Vantage, Poast

Comments: Systemic, selective, water-soluble herbicide used to control annual and perennial grasses. Low persistence in the soil. Does not control annual or perennial sedges, annual bluegrass, or broadleaf herbs and woody plants. Do not use in areas of standing water. Often best to apply when native grasses are dormant and non-native grasses are vulnerable. Rapid degradation can limit effectiveness. Relatively low toxicity to mammals, birds, and aquatic animals. It can be highly mobile because it is water-soluble and does not bind strongly with soils. Use of some products may be restricted in some states.

Triclopyr

General comments: Available in a variety of formulations. All have soil residual activity and only moderately bind with soil particles. Off-site movement through runoff is a possibility. Relatively nontoxic to terrestrial animals.

Trade name: Brush-B-Gon

Comments: Water-soluble, amine formulation. Can be used as a foliar spray or for cut-stump treatment on vines, climbing shrubs, and small trees whose green tissue is exposed when the bark is scratched with a fingernail. Can also be used for cut-stem treatment of herbaceous plants. May harm grasses due to strength of formulation. Available in one-quart containers.

Trade name: Crossbow (ester formulation, also contains 2, 4-D ester)

Comments: Systemic formulation for the control of most woody and herbaceous broadleaf weeds in noncrop areas. Can be mixed with water or oil. May be available in smaller quantities than other triclopyr products.

Trades names: Garlon 3A, Renovate

Comments: Broadleaf-specific, systemic, water-soluble, amine formulation. Can be used as either a foliar spray or for cut-stump treatment. Apply to cambium layer of cut stump. Do not apply down sides of trunk, at temperatures below freezing, or during spring sap flow. A surfactant should be included for foliar treatments. Highly alkaline and can severely damage eye tissue. Renovate is formulated and approved for use in and near bodies of water. Only available in two-and-a-half-gallon containers and larger.

Trade name: Garlon 4

Comments: Broadleaf-specific, systemic, ester formulation. Can be mixed with bark-penetrating oil or water for cut-stump treatment. When using water for cut-stump treatment, apply only to cambium layer. Do not apply down sides of trunk, at temperatures below freezing, or during spring sap flow. If using penetrating oil for cut-stump treatment, apply to cambium layer and down sides of trunk. Must be mixed with bark penetrating oil for basal bark treatments. Performance of basal bark treatment is inconsistent on stems greater than six inches in diameter. Foliar sprays must be mixed with water and a surfactant. Vapor drift possible. Best applied on days with cool temperatures. Highly toxic to aquatic organisms. Available only in two-and-a-half gallon containers and larger.

Trade name: Pathfinder II

Comments: Ready-to-use, but volatile, ester formulation for the control of woody plants by either basal bark or cut-stump treatment (apply to cambium layer and down sides of trunk). Basal stems should be less than six inches in diameter for consistent results using basal bark treatment. Use on noncropland areas. Available only in two-and-a-half-gallon containers and larger.

Trade name: Vine-X

Comments: Ready-to-use mixture of triclopyr ester and penetrating oil for the control of woody plants by simply painting the herbicide on the bark of vines or the cut stumps of small trees and shrubs. Brush applicator is built into the container. Available in two sizes: twelve-ounce bottle with a 0.125″-wide brush and a sixteen-ounce bottle with a 0.375″-wide brush applicator. www.vine-x.com.

2,4-D

Trade names: Amine 2,4-D, See 2,4-D LV4 (ester), and many more

Comments: Systemic, plant growth regulator that is available in both amine (dimethylamine salt of 2,4-D) and ester (butoxyethyl ester or isooctyl ester) formulations. (Check active ingredients on the label.) Only amines are water soluble. Amine formulations are not volatile and are better for use in warmer months when volatile ester formulations are more likely to damage nontarget plants. Typically applied to the foliage, it is used on broadleaf herbaceous and woody plants. Grasses are not harmed if used according to label directions. Has no soil residual activity. 2,4-D is generally slightly to moderately toxic to animals, but some formulations are highly toxic to fish and aquatic invertebrates, and some can cause severe eye damage. Other formulations are registered for use against aquatic weeds. 2,4-D is readily absorbed through skin and lungs. Carcinogenic effects are unclear. Relatively inexpensive when compared to some other herbicides.

Purchasing Herbicides

The variety of trade names can make it confusing when trying to locate and purchase a particular chemical. Check herbicide labels under the "Active Ingredients" section for the chemical names. Common names will not be given. The percentage of active ingredient (a.i.) listed may vary for any one chemical and will determine the strength of the herbicide compared to other formulations and its appropriate use. Once again, always follow label directions.

Residential retailers, such as hardware stores and garden centers, may not carry some herbicides. Agricultural retailers or co-ops often have them. Check the Yellow Pages or the Internet under "Fertilizers," "Chemical Dealers," "Herbicides," or "Garden Supplies," to locate a business near you.

Herbicide Information

- Crop Data Management Systems, Inc. (www.cdms.net/), includes label and MSDS information.
- Extoxnet (ace.orst.edu/info/extoxnet/), a cooperative effort by five universities provides objective, science-based information about pesticides, including herbicides. The information is written for the nonexpert and includes toxicology information trade names, toxicity ratings, reproductive and carcinogenic effects, impact on other organisms, and persistence in the environment.
- The Nature Conservancy: tncweeds.ucdavis.edu/handbook.html
- National Pesticide Information Network: (800) 858-7378; npic.orst.edu
- Chemical Referral Center: (800) 262-8200
- Environmental Protection Agency, Office of Pesticide Programs: www.epa.gov/pesticides/

Biological Control

Biological control methods use animals, fungi, or microbial diseases to control a particular species. When invasive plants are considered, scientists usually look for biological controls from the targeted plant's native range. Because newly introduced biological controls could harm native vegetation, crops, and important ornamental species, scientists thoroughly study a variety of control agents to make sure that they will remain specifically associated with the invasive plant that needs control.

Although biological controls will not eliminate a targeted invasive plant and may take several years to become successful, some biocontrol efforts have proven very effective. These cases include the biocontrol of Klamathweed (*Hypericum perforatum*) and tansy ragwort (*Senecio jacobaea*) in the Pacific Northwest, and alligatorweed (*Alternathera philoxeroides*) in the South and Southeast. Biocontrol efforts aimed at reducing populations of purple loosestrife (*Lythrum salicaria*) and leafy spurge (*Euphorbia esula*) also show great promise. Biocontrol agents have failed, however, to control Canada thistle (*Cirsium arvense*), and spotted and diffuse knapweed (*Centaurea maculosa* and *C. diffusa*). There have also been a few problems, especially early in the development of biocontrol technology, with biocontrol agents eating desirable native species.

Despite failures, researchers continue to assess and test biological controls for many invasive plants. At the present time, researchers and land managers have released biocontrol agents for about seventy invasive plant species in the continental United States and Hawaii. Research is underway to determine if biocontrol agents can safely and effectively control about twenty-five more invasive species, including garlic mustard (*Allaria petiolata*) and the invasive buckthorns (*Rhamnus cathartica* and *R. frangula*). These often lengthy efforts are worthwhile because biological controls are:

- relatively inexpensive when compared to the cost of human labor or chemicals needed to control invasive weeds, especially in large, weed-infested areas;
- usually less harmful to the environment than many chemical controls;
- able to spread naturally and control a larger area than where they were initially introduced;
- able to remain active indefinitely or until the food supply of invasive plants is exhausted.

It is best to obtain any biological control agents locally. Check with your state's agriculture agency or with the USDA-APHIS Plant Protection and Quarantine Office in your state before obtaining or releasing biocontrol agents reared in another state. For more information on biocontrol, see www.nysaes.cornell.edu/ent/biocontrol/ and www.invasive.org/biocontrol.cfm.

It is extremely important to monitor the effectiveness of biological control agents. Each release site should be mapped, and before/after photographs should be taken at the same spot, same time of day, and with the same or similar equipment.

Grazing

Animal grazing can either promote or reduce invasive plant infestations at a particular site. It can be especially helpful in areas where herbicides cannot be used, such as near water or when invasive plant infestations are so extensive that herbicide use would be too expensive. Overgrazing or grazing at the wrong time, however, can disturb the soil, reduce and trample native plant populations, spread weed seeds, and increase the potential for weed invasions. Cattle, sheep, goats, horses, and geese are among the animals most frequently used as part of a invasive plant control program. A grazing plan should be developed to determine the best species of grazers to use, and the timing, duration, and frequency of grazing.

Points to remember when using grazing as a form of invasive plant control:

- It is most effective when used in combination with some other form of invasive plant control, such herbicides or biocontrol.
- Grazing should be timed (e.g., during flower and seed production) to cause the maximum damage to the invasive plants and the least amount of damage to native vegetation. In some cases, for example, grazing can take place before native plants emerge in the spring. Remember that some plants may be unpalatable to livestock at certain stages of the growing season.
- Cattle and geese may help control invasive grasses.

- Horses are more choosey about the plants they eat than are cattle.
- Sheep and goats prefer to eat broadleaf herbaceous plants. They have been used to control leafy spurge (*Euphorbia esula*), Russian knapweed (*Centaurea repens*), and toadflax (*Linaria* spp.), and seem to neutralize the toxins present in some forbs better than other animals. Sheep can also be used to control spotted knapweed (*Centaurea maculosa*), oxeye daisy (*Chrysanthemum leucanthemum*), and kudzu (*Pueraria montana*), but should not be used to control St. John's wort (*Hypericum perforatum*) because of its toxicity. Goats will eat a larger variety of plants than sheep and can stand on their hind legs to eat woody vegetation that other animals cannot reach.
- Grazing animals can pass viable seeds through their digestive tracts and should not be allowed to enter uninfested areas for five to nine days after eating weed seeds. Remember also that seeds can be transported to new areas on their fur or hooves.
- Be aware of animals in the area that can prey on certain grazing animals. This may help determine which animals are selected for grazing purposes.
- Some animals take time to adjust to a new food in their diet. When using sheep, it is often helpful to have animals that are experienced in eating certain weeds to lead the way for other sheep.
- Fencing, water, and salt licks may be necessary to help concentrate grazing in a particular area.

(Based on information from The Nature Conservancy's "Weed Control Methods Handbook," Tu et al., April 2001 version, tncweeds.ucdavis.edu/handbook .html.)

Weeding Tools

Root Talon

The Root Talon is a multipurpose tool that consists of a fiberglass shaft to which is attached a metal hoelike head on one side that can be used to grub out herbaceous and small woody plants. The other side of the toolhead can be used to grip the base of a herbaceous plant, small shrub, or tree, and leverage it out of the ground. Root Talons are lightweight and relatively inexpensive. For more information, contact Lampe De-

sign, LLC, 262 South Griggs Street, St. Paul, MN 55105, (612) 699-4963, jklampe@worldnet.att.net or www .lampedesign.com/.

Weed Wrench

The Weed Wrench is similar to the Root Talon in that it can be used to pull plants out of the ground, but it can handle larger plants and provides greater leverage and pulling power. The Weed Wrench is available in four sizes—heavy, medium, light, and mini. Larger sizes can be heavy and cumbersome in the field. For more information, contact New Tribe, Tom Ness and Sophia Sparks, 5517 Riverbanks Road, Grants Pass, OR 97527, (514) 976-9492, Newtribe@cdsnet.net or www.canonbal.org/ weed.html.

Girdling Tools

Although axes and saws can be used to girdle woody species (see "Controlling Woody Plants," page 13), simple tools, like the Ringer Tree Girdler, are available to make this job easier. Forestry or gardening supply catalogs often carry them or call Forestry Suppliers, Inc., (800) 647-5368, www.forestry-suppliers.com. You can also visit the Forestry Mall Web site, www.irl.bc .ca/Forestry%20Supplies/axes-tools-2.htm. Lampe Designs, the manufacturer of the Root Talon, sells an easy-to-use and effective girdling tool known as the Weed Tree Eradicator.

Herbicide Wand Applicators

Herbicide wand applicators are used to apply herbicides to foliage or cut stumps. They typically consist of an herbicide reservoir connected to either a small or a long, narrow sponge by plastic or PVC tubing. Most applicators have a valve that can be adjusted to evenly control how much herbicide reaches the sponge. Wand applicators are more selective in distribution of herbicide to foliage than spray applications because they virtually eliminate drip and spray drift. For instructions on how to build your own for cut-stump treatments, visit tncweeds.ucdavis.edu/tools/wand.html. Commercially available products include the Weed Wand and Vine-X.

A version of a wand-type applicator is the Hypo-Hatchet Tree Injector. This tool has a hatchet instead of a sponge at the end of the PVC tubing. The operator strikes the unwanted tree trunk at a forty-five-degree angle with the hatchet head, and the force of the blow is enough to deposit a preset amount of herbicide into the cambium layer. The tool works only with amine herbicides, such as triclopyr and amine 2,4-D. Forestry Suppliers and other land management supply firms carry this product.

Chainsaws

Chainsaws are helpful in removing trees or shrubs that are too large to remove with a Root Talon, Weed Wrench, or hand saw. Many sizes are available with electric or gas-powered motors. Chainsaws are dangerous to operate, however, and should never be used without training and unless one other person is nearby. For more information, visit www.stihlusa.com/manuals/chainsaw_safety_english.plf.

Brush Cutters and Brush Mowers

Brush cutters are often used to remove dense, brushy vegetation in areas primarily comprised of tree saplings and small shrubs. They are usually easier and faster to use than chainsaws, hand saws or handheld pruners in these areas. Available in a variety of styles and sizes, brush cutters can range from handheld cutters to walk-behind mowers. Handheld cutters often come with shoulder harnesses and a chisel-tooth, circular-type blade mounted on a long arm. They can cut through individual woody stems up to about four inches in diameter. Walk-behind mowers are used to clear a path, cutting multiple stems simultaneously. Stems can be up to about three inches in diameter, depending on the model. Cutting can reduce seed production and restrict weed growth but may also lead to vigorous resprouting unless cut stumps are treated with herbicide. Handheld brush cutters make cleaner cuts than walk-behind models, making herbicide application more effective. Cut fragments of some plants may need to be collected and removed if they are prone to resprouting from cut pieces.

Large brush mowers, like the AcrEase from Kunz Engineering (www.kunzeng.com), are available for pulling behind a tractor or an ATV. Some models, like the Brown Brush Mower, include an herbicide applicator to treat cut stems and stumps.

Invasive Plants of Major Concern

THE PLANTS DISCUSSED in this chapter are causing serious ecological and economic damage in significant areas of the upper Midwest. Although they may not currently be a problem everywhere in the region, they have clearly demonstrated an ability to spread aggressively and cause considerable environmental disruption. They should be taken seriously.

TREES AND SHRUBS

Autumn Olive
Elaeagnus umbellata

Autumn olive is a deciduous shrub or small tree in the Oleaster family that is native to China, Japan, and Korea. Introduced to the United States in 1830, it has been widely used in landscaping and for wildlife food and cover. Autumn olive has also been planted for screening along highways, to stabilize and revegetate road banks, and for strip mine reclamation. Once established, this highly invasive and difficult to control species can eliminate almost all other plant species. Because autumn olive fixes nitrogen in the soil, it can disrupt native plant communities that require infertile soil.

Autumn olive is a large shrub or small tree.

Leaves are simple with entire, somewhat wavy margins; flowers fade after initial bloom.

Initially the flowers are light yellow.

> *Habitat:* Disturbed areas, roadsides, open woods, forest edges, fencerows, prairies, pastures, sand dunes; can adapt to poor soils; rarely found in dense forests or on wet sites.
>
> *Height:* 6–20′.
>
> *Leaves:* Simple; alternate; small; elliptical or oval; margins untoothed but often slightly wavy; upper surface is dark green to grayish green, silvery beneath due to white scales—a conspicuous trait that can be seen from a distance; produces leaves in early spring while most native vegetation is still dormant.
>
> *Flowers:* Light yellow; tubular, arise from leaf axils along twigs; bloom in May or June after the first leaves appear, fragrant.
>
> *Fruit:* Small; round; pink to red with silvery or coppery spots; abundant—individual plants can produce up to eight pounds of fruit in one season; birds and falling fruit are the primary means of seed dispersal; also eaten by raccoons, skunks, and opossums; seeds germinate readily.
>
> *Stems:* Have coppery dots; may have thorns that are several inches long.
>
> *Similar species:* Resembles Russian olive (*Elaeagnus angustifolia*), another invasive exotic species (see chapter 4). Russian olive leaves are more silvery, longer, and narrower. Leaves of both species are silvery beneath. Silver buffaloberry (*Shepardia argentea*), which is found in far western Iowa and Minnesota, and russet buffaloberry (*S. canadensis*), which occurs along the margins of the Great Lakes, are native shrubs that look somewhat similar to autumn olive. Unlike autumn olive, however, they have opposite leaves that are silvery on both sides.

Fruits are reddish with silver or brown spots.

Underside of leaves are silvery.

widespread in the upper Midwest. Pockets of infestation have also resulted after they were planted to "improve wildlife habitat" or to control erosion. First introduced to North America from Eurasia in 1752, honeysuckles' vigorous growth and early spring leaf out inhibit the growth of native shrubs and groundlayer species, reducing food and cover for wildlife. If not controlled, they can become massive in size and eventually replace native plants in a given area by crowding or shading them out and by depleting the soil of moisture and nutrients. Some studies suggest that exotic honeysuckles may also be allelopathic. Although Bella's, Morrow's, and Tartarian varieties are well established across vast areas of the upper Midwest, Amur honeysuckle is

Control:

Manual or Mechanical Control: Young plants and seedlings can be hand pulled in early spring when there is adequate ground moisture. The entire root system must be removed to prevent regrowth. Burned, mowed, or cut plants will resprout vigorously unless herbicide is applied to the stump.

Chemical Control: Cut-stump treatments using 20–25% active ingredient (a.i.) glyphosate or 12.5% a.i. triclopyr formulated for use with penetrating oil have been effective. Use a low-pressure hand sprayer, wick, or sponge paintbrush to apply. If resprouts occur after cut-stump treatment, cut and treat each stem again, or carefully apply 5% a.i. glyphosate to the foliage of resprouts. A 12.5% a.i. solution of triclopyr formulated for use with penetrating oil is also effective as a basal bark treatment on this species.

A broadleaf specific solution of metsulfuron-methyl plus a surfactant can also be used as a foliar treatment. This mixture will have preemergent control of germinating seedlings.

[Based on information obtained from the Illinois Nature Preserves Commission.]

Exotic Honeysuckles

Bell's Honeysuckle: *Lonicera* x *bella*
Amur Honeysuckle: *Lonicera maackii*
Morrow's Honeysuckle: *Lonicera morrowii*
Tartarian Honeysuckle: *Lonicera tatarica*

Exotic honeysuckles are upright deciduous shrubs that have spread from horticultural plantings and are now

Honeysuckle

In a six-year study at the Morton Arboretum near Chicago, researchers found that robins and wood thrushes that nested in exotic buckthorn and honeysuckle shrubs suffered much greater predation than those that nested in comparable native shrubs like hawthorns and viburnums. Researchers believe that nests placed in these non-native shrubs are more vulnerable to predators compared to nests placed in the native species. For example, the stronger limbs of honeysuckle encourage birds to build nests lower than they often do in native shrubs, making it easier for raccoons and other marauders to prey on them. In addition, neither honeysuckle nor buckthorn have stout thorns like the hawthorns to protect the birds from mammalian predators. Because both non-native shrubs leaf out earlier than native shrubs, the birds are also more likely to choose these "ecological traps" for nesting early in the season. (Kenneth A. Schmidt and Christopher J. Whelan, "Effects of Exotic *Lonicera* and *Rhamnus* on Songbird Nest Predation," *Conservation Biology,* December 1999)

Note: Although not implicated by the study above, cats, also non-native in origin, are similarly deterred as predators by thorny plants like hawthorns. Research by John Coleman, Stanley Temple, and Scott Craven indicates that in Wisconsin alone rural free-ranging domestic cats kill about 39 million birds per year.

Cedar waxwings, gorgeous birds with striking black face masks, normally have a bright yellow band across the tip of their tails, but over the last thirty-five years, an increasing number of waxwings have been sporting an orange tail band instead. Laboratory studies in 1992 in the vicinity of Ithaca, New York, demonstrated that yellow-tailed waxwings, which were fed honeysuckle berries (*Lonicera morrowii*) during molt, grew orange tails. Bird banders have also found orange replacing yellow in some white-throated sparrows, Kentucky warblers, and yellow-breasted chats. Because color is so important to the social behavior of birds (e.g., mate selection), this introduced shrub could have far-reaching and adverse effects for all of these birds. (Based on information supplied by Mark Witmer, Section of Ecology and Systematics, Cornell University, Ithaca, NY)

Exotic honeysuckles are among the first shrubs to produce leaves in spring.

Leaves are opposite with smooth edges.

Honeysuckles are frequently used for landscaping.

Flowers are paired and vary in color from white to pink to crimson, depending on the species or cultivar.

Berries are paired and may be red, orange, or yellow.

currently most problematic in the southern areas of this region but is spreading northward. Bella honeysuckle is a hybrid of Morrow's and Tartarian honeysuckle.

Habitat: Thrive in sunny, upland sites including forest edges, roadsides, pastures, and abandoned fields; also found in fens, bogs, and on lakeshores; relatively shade intolerant. While most natural communities are susceptible to invasion, open woods are most affected and particularly vulnerable if the site has been disturbed.

Height: 6–18′, although height varies among species.
Amur: to 18′.
Bella's: to 18′.
Morrow's: to 6′.
Tartarian: to 9′.

Leaves: Simple; opposite; entire; oval, oblong, or elliptic; short petioles. Exotic bush honeysuckles are easy to spot at the beginning and end of the growing season. In spring they normally produce leaves one or two weeks before native trees and shrubs. This is especially harmful to spring wildflowers since they have evolved to bloom before native trees and shrubs have leafed out. Exotic honeysuckles also hold their leaves later in the fall.
Amur: Elliptical; slightly hairy; 1.5–3.5″ long; the only exotic bush honeysuckle whose leaves taper to a long, narrow point.
Bella's: Elliptic to oblong or oval; slightly hairy beneath; 1–2.5″ long.

Bark is gray, rough; twigs are often hollow.

Morrow's: Elliptic to oblong; short, pointed tips; gray-green; softly hairy beneath; 1–2.5″ long.

Tartarian: Oval to oblong; short, pointed tips; hairless; 1–2.5″ long; bluish green.

Flowers: Abundant, pink to purplish red or white fading to yellow with age, tubular, arranged in pairs arising from leaf axils; bloom in May and June; fragrant.

Amur: White, sometimes tinged with pink, yellow with age; peduncles (stems supporting flower clusters) are 0.2″ long or less.

Bella's: Pink, fading to yellow with age; slightly hairy peduncles are 0.2–0.6″ long.

Morrow's: White, yellow with age; hairy; densely hairy peduncles are 0.2–0.6″ long.

Tartarian: Usually pink to purplish red, sometimes white; hairless; peduncles are 0.6–1″ long.

Fruit: Abundant; red, although Amur may be orangish, paired berries; ripen in early summer; birds readily eat berries and disperse seeds over great distances; seeds are viable two to three years.

Stems: Multiple; many branches; arching; sometimes develop roots where they touch the ground.

Bark: Gray or tan; shaggy; older branches are often hollow.

Similar species: Exotic bush honeysuckles are easy to distinguish from native *Lonicera* species. While the exotic honeysuckles are stout, erect shrubs, the native honeysuckles are either woody, vinelike species as is the case with red honeysuckle (*L. dioica*), yellow honeysuckle (*L. flava*), hairy honeysuckle (*L. hirsuta*), and grape honeysuckle (*L. reticulata*), or fairly short, sparse shrubs like Canada honeysuckle (*L. canadensis*), American fly honeysuckle (*L. involucrata*), fly honeysuckle (*L. oblongifolia*), and swamp fly honeysuckle (*L. villosa*).

Species of the genus *Diervilla* are native bush honeysuckles with yellow flowers and bronze to green leaves. They grow about four feet tall and are often found on dry or rocky sites.

Control: Control methods must be repeated for at least three to five years to deplete the seedbank.

Manual or Mechanical Control: Since honeysuckle roots are fairly shallow, small- to medium-sized plants can often be dug or pulled out by hand or with a leverage tool (see "Weeding Tools," in chapter 2, page 29). This is especially easy in the spring when the soil is moist.

Cut honeysuckle will resprout vigorously if the stump is not treated with an herbicide (see Chemical Control). In small areas, such as urban yards, honeysuckle can be controlled without herbicides by cutting several times a year for at least two years until the plant's root stores are depleted and it dies.

The pulled plants or cut stems of honeysuckle sometimes reroot if they remain in contact with the soil. Placing smaller pulled plants or cut stems on nearby bushes or tree branches to dry them out will help prevent this. When working with large plants or numerous cut stems, plants can be placed in piles for burning or easier monitoring.

In fire-adapted plant communities, controlled burns in spring may kill honeysuckle seedlings and top-kill larger plants. Resprouts may appear, so repeated burning for several years may be necessary.

In areas where honeysuckles have been removed, replanting with native species appropriate for the site will discourage reinvasion. In sensitive areas, physical removal of honeysuckles and soil disturbance should be avoided if it is likely to harm native plant communities. Appropriate chemical control may be less damaging in these areas.

Chemical Control: Honeysuckles hold their leaves later in the fall than native shrubs. This makes them easy to spot for control treatments at this time. Cut-stump treatments using 20% a.i. glyphosate or 12.5% a.i. triclopyr formulated for use with penetrating oil have been effective on honeysuckles. Triclopyr with oil can be used throughout the year and is most successful when applied in the winter, least successful when applied in spring. If stumps cannot be treated at the time of cutting, resprouts can be sprayed with 5% a.i. glyphosate. Cut-stump treatment using triclopyr formulated for water dilution has not been effective on this species.

Basal bark treatment using 12.5% a.i. triclopyr formulated for use with penetrating oil is also effective on honeysuckles.

If using a foliar spray, which is less selective than other methods of herbicide applications, spray new foliage in the spring before the leaves of native shrubs and ground flora emerge. Metsulfuron-methyl plus a surfactant is broadleaf specific and can be used as a foliar spray. If using glyphosate, which is nonselective, a 1.5% a.i. solution should be effective.

Non-Native Buckthorns
Common buckthorn: *Rhamnus cathartica*
Glossy buckthorn: *Rhamnus frangula,* syn. *Frangula alnus*

Common and glossy buckthorn, two closely related species, grow in the upper Midwest as either small trees or tall shrubs. Originally from Eurasia, they were

Common buckthorn is a small tree with few to several trunks.

Narrow, upright buckthorn cultivars are often used for hedges.

Common Buckthorn

The monocultures formed by invasive exotic plants are particularly unfit for bird habitat. Birds require habitat containing a large number of plant species, variation in horizontal canopy cover, and complexity in vertical structure. By contrast, invasive exotic infestations are composed primarily of one species, mostly uniform in height and structure, forming a canopy coverage that is too dense. (Jerry Asher, Bureau of Land Management, "The War on Weeds: Winning It for Wildlife," 2000)

Glossy buckthorn

In just 20 years, glossy buckthorn has overrun one-third of the 2,500-acre Cedarburg Bog, a DNR and UW-Milwaukee preserve that is unique in Wisconsin. Our budget will pay for eradication on only 19 acres. (Jim Reinartz, University of Wisconsin–Milwaukee Field Station, 2001)

In a study done in the spring of 2000 by the Zoological Society of Milwaukee, fewer arthropods were found on common and glossy buckthorn than on eleven species of native trees and shrubs. Thirty-two samples of red oak branch clippings, for example, contained a total of 328 arthropods while the same number of common buckthorn clippings had only 58 arthropods. (Victoria D. Piaskowski, Birds without Borders-*Aves Sin Fronteras*, International Coordinator)

introduced to the Midwest as ornamentals as early as 1849. They are now well established and spreading rapidly. Although common buckthorn is more widespread, glossy buckthorn is particularly aggressive in wet areas. Cultivars of glossy buckthorn are frequently planted for visual screening and in areas where space is too limited for larger trees. Both species produce a dense shade that eliminates native tree seedlings, saplings, and groundlayer species. As a result, the ability of forests to regenerate and remain healthy can be severely limited as buckthorns multiply. Both species have:

- long growing seasons—they leaf out early and hold leaves late into the fall. Because they remain green well after most native trees and shrubs have lost their leaves, they are easy to spot at this time;

Buckthorn often shade out native plants growing underneath them.

Multitrunk glossy buckthorn often grows in low-lying areas.

- prolific production of berries that are attractive to birds who spread the seeds;
- a rapid growth rate and the ability to grow in a wide variety of habitats, including nutrient-poor soils, full sun and dense shade, and wet soils.

Several states, including Minnesota and Illinois, have laws preventing the sale of non-native buckthorns and their cultivars.

Other common names:

Common buckthorn: European buckthorn, Hart's thorn, European waythorn, Rhineberry

Glossy buckthorn: Alder, Tallhedge, Columnar, Fernleaf, and Fen buckthorn; European alder

Habitat:

Common buckthorn: Prefers woods or woodland edges; is particularly aggressive on well-drained soils; also found in prairies, thickets, old fields, yards, and along roadsides.

Glossy buckthorn: Woodland edges, full sun to heavy shade; fens; most aggressive in wet soil but also found in drier areas.

Height and Growth Form: Both species can reach 10–25′ in height; trunks can be up to 10″ in diameter; few to several stems at the base and spreading, loosely branched crowns. Cultivars of glossy buckthorn generally have a more narrow, upright form.

Leaves:

Common buckthorn: Simple; mostly opposite; dark green; shiny on upper surface; hairless on both sides; ovate to elliptic; minute teeth on margins; rounded to pointed at the tips; veins curve toward leaf tips.

Glossy buckthorn: Simple; mostly alternate; thin; ovate to elliptic; upper

A buckthorn seedling.

Glossy buckthorn leaves are mostly alternate unlike the opposite leaves of common buckthorn.

surface is shiny; lower surface is duller and can be hairy or smooth; margins are untoothed; veins (five to ten) extend straight out from midrib then turn toward tip near leaf margins.

Thorns:

   Common buckthorn: Often found on tips of twigs.

   Glossy buckthorn: Absent.

Flowers:

   Common buckthorn: Small, greenish yellow, four petals, clustered in leaf axils near ends of stems; bloom May through June, fragrant.

   Glossy buckthorn: Small, pale yellow, five petals, one to eight clustered in leaf axils; bloom late May to first frost.

Fruit:

   Common buckthorn: (On female plants only) Black, rounded, pea-size, clustered; ripen August through September; may remain on the tree until the following spring; have a severe laxative affect on birds that eat them resulting in loss of energy; contain three to four seeds that remain viable in the soil for two to three years.

   Glossy buckthorn: Progressively ripen from a distinctive red to dark purple, pea-size; develop early July through September; seeds remain viable in the soil for two to three years; have laxative affect on birds.

Bark:

   Common buckthorn: Brown or gray when young; becomes gray-black, rough and peeling with age; has prominent, often elongated, light-colored lenticels running parallel to twigs, lenticels are especially noticeable on young plants. Note: Native plums and cherries have similar bark.

   Glossy buckthorn: Gray or brown with prominent, closely spaced, often elongated, light-colored lenticels.

Sapwood: Yellow in both species.

Thorns often occur on twigs; young bark has white spots.

Female plants produce clusters of black berries.

Fruits ripen from red to dark purple and have a cathartic effect on birds.

Heartwood: Pinkish to orange in both species.

Terminal buds:

Common buckthorn: In pairs that resemble a buck's hoof with a thorn frequently protruding between the two buds, hence the name "buckthorn." (Other buds on twigs are also usually or nearly opposite.)

Glossy buckthorn: Rust colored, hairy.

Similar species: Two small native shrubs, which are not invasive, resemble exotic buckthorns. Alder buckthorn (*Rhamnus alnifolia*) grows less than 3′ in height and has hairless twigs and dark scales on the buds in winter. Lance-leafed

Buckthorn bark becomes rough with age.

Bark has elongated lenticels.

Terminal buds resemble a buck's hoof and frequently have a thorn protruding from its center, hence the name "buckthorn."

buckthorn (*R. lanceolata*) grows less than 6′ in height and is found in bogs and swamps. Its alternate leaves are 2–6″ in length and taper to a point at the tip. Bud scales can be seen on twigs in winter.

Control:

Manual or Mechanical Control: The most effective control for buckthorns is to recognize their presence early and remove plants before they begin to produce fruits. Small plants can be pulled when the soil is damp. Larger plants, those 0.5–1.5″ in diameter, can be dug or pulled by using a leverage tool (see chapter 2, "Weeding Tools").

When a large number of buckthorn seedlings are present, controlled burns in fall or early spring in fire-adapted plant communities may kill seedlings, especially those in the first year of growth. Burning may need to be repeated annually or biannually for two or three years to deplete the seed bank. Such frequent burning should not be used if it will adversely affect the native plant community. Burning dense buckthorn stands is generally difficult since the understory is typically well shaded, allowing little fuel (dead grasses, sedges, and leaf litter) to accumulate. In areas that lack sufficient fuel, it may be necessary to use a propane torch to burn the seedlings.

Girdling has produced mixed results for controlling buckthorn. This method involves removing a 2″ band of bark from each stem arising from the roots. For best results, apply herbicide to the girdled surface.

In wetlands with artificially lowered water tables, restoring water to its previous levels will often kill glossy buckthorn in the area.

Hairy, rusty-colored terminal buds help with winter identification.

Chemical Control: (Note: In areas of standing water, use herbicides approved for use in aquatic sites. Permits may be required.)

Late fall is an ideal time for chemical control of buckthorn stems and trunks because most native plants are dormant at this time, and chemicals are easily drawn down to the roots with natural sap flow. Buckthorns are also easy to spot in the landscape this time of year because they typically hold their green leaves two to three weeks longer than native trees and shrubs. For large infestations, eliminate the largest, seed-producing plants first.

Cut-stump treatments using 20–25% a.i. glyphosate or 12.5 % a.i. triclopyr formulated for dilution with oil have been effective. Use a low-pressure hand sprayer, wick, or sponge paint brush to apply. If resprouts occur after cut-stump treatment, cut and treat each stem again, or carefully spray or wick 1.5% a.i. glyphosate on the foliage of resprouts.

For buckthorn plants smaller than 6″ in diameter at the base, basal bark applications of 12.5% a.i. triclopyr formulated for dilution with oil have been effective. A solution of 6% a.i. triclopyr has been effective on plants that are less than 2.75″ in diameter at 18″ above the ground. [S. Glass] Stems larger than 2″ in diameter should be sprayed entirely around the stem. For smaller stems, spraying one side is sufficient. Basal bark treatment is inconsistent on trees larger than 6″ in diameter.

In badly infested areas, buckthorn seedlings can be controlled by applying 1.5% a.i. glyphosate or a 1% a.i solution of triclopyr with water to the foliage using a long-handled wick or low-pressure spray. To help avoid herbicide contact with native plants and depending on the site, seedlings can be treated in the fall after a hard frost when most native plants have gone dormant. Desirable nonwoody, native perennials in the area that have not yet gone dormant can be cut back to ground level before herbicide application to help limit their exposure.

Fosamine, a bud inhibitor for woody species, is effective when applied to buckthorn foliage in September. Areas of the plant where herbicide contact was made will not leaf out the following spring.

Biological Control: Researchers are studying the potential use of several insects from Europe as biological control agents for non-native buckthorns. Preliminary indications lead them to believe there may be several useful insects, but they will require further extensive testing before any large-scale releases are made. The earliest releases of these insects in North America will likely be made between 2007 and 2010.

Black Locust
Robinia pseudoacacia

Black locust is a translocated, deciduous tree of the Legume family. Native to the slopes and forest margins of southern Appalachia and the Ozarks, it was planted extensively in the upper Midwest in the early 1900s to prevent soil erosion. Black locust seeds were commonly planted to help create windbreaks during the Dust Bowl era of the 1930s. Besides providing nectar for bees, black locust wood burns well and is valued for its durability and use for fence posts and shipbuilding. In natural areas, however, dense stands can develop and shade out important native vegetation. Black locust is also suspected of producing chemical compounds that prevent other plants from growing. Leaves, seeds, and bark are toxic to livestock and humans if ingested.

Habitat: Forests, especially in hilly areas, pastures, old fields, fencerows, and roadsides; also found in upland

Black locusts are large trees that bloom in May or June.

prairies and savannas; prefers rocky, sandy or loamy, well-drained soil in open, sunny locations.

Height and Growth Form: 30–90′; 2–4′ diameter trunk; clonal.

Leaves: Pinnately compound; alternate; up to 12″ long, 7–21 paired leaflets with an additional leaflet at the tip, leaflets entire, thin, oval, dark blue-green above and pale beneath; slow to leaf out in spring making plants easy to identify at that time.

Thorns: Paired; heavy; 0.25–1″ long; found on smaller branches where leaves attach to stem.

Paired thorns are found on smaller branches.

Flowers: White and yellow, pealike, borne in large, showy drooping cluster 4–8″ long; bloom in May and June, fragrant.

Seedpods: Shiny; smooth; narrow; flat in cross-section; 2–6″ long; dark brown when ripe; contain four to eight seeds with thick coats that often prevent germination; develop in September and most remain on trees to the following spring.

Bark: Young saplings have smooth, green bark; older trees have dark bark with deep furrows and flat-topped ridges.

Root system: Extensive; spreads by shallow rhizomes; forms massive clones by root suckering and stump

Pealike, white flowers occur in large, drooping clusters.

Distinctive long, flat seedpods develop by midsummer.

sprouting; roots are fibrous; oldest trees are located in the center of a clone and youngest trees are on the outside; damage to roots or stems from fire, wind, cutting, or disease stimulates vigorous sprouting, root suckering, and lateral spread.

Natural plant enemies: Black locust is susceptible to locust borers, locust leaf miners, and locust twig borers that cause deformed growth and dieback.

Reproduction: By seed, shoots rising from rhizomes of established plants, and stump sprouting.

Control:

Manual or Mechanical Control: Cutting black locust without applying herbicide will stimulate sprouting and clonal spread. Mowing and burning black locust will temporarily control this tree, but mowing also seems to promote seed germination, while burning stimulates sprouting. Girdling is ineffective because it kills the stem but does not prevent sucker formation. Annual mowing may be enough to control first-year seedlings and prevent black locust from spreading into prairie communities. Bulldozing may be an option on highly disturbed sites.

Chemical Control: When using chemicals to control black locust, all stems (trees) in a clone must be treated.

A basal bark application of 12.5% a.i. triclopyr formulated for use with penetrating oil can be used on black locust, applied directly to the base of each stem in a broad ring. The triclopyr/oil mixture can also be applied to a 2–3″ girdle cut made with a chain saw at 3.5′ from the ground or to cut stumps. In the latter case, apply triclopyr to the cambium layer and down the sides of the trunk.

A solution of 20% a.i. glyphosate can be applied to the cambium layer of black locust stumps immediately after cutting. Cut-stump treatment with glyphosate works best when applied in late summer, early fall, or during the dor-

Bark of a mature tree.

mant season when temperatures are above freezing. This technique works better at some sites than others because cutting may stimulate sprouting.

Picloram and dicamba (formulations for use on cut stumps) have also been used successfully as cut-stump or girdle treatments for black locust.

For small, isolated plants or thick patches under 5′ in height (such as those resulting from cutting or fire), fosamine can be applied as a foliar spray from July to September 15. It kills plants by inhibiting leaf bud growth and flower formation the following spring. In order to effectively curb regeneration, every branch or stem must be sprayed because missed stems will leaf out. Fosamine can also be applied as a cut surface treatment using 50% water and 50% fosamine. Triclopyr (1 to 2% a.i.) formulated for mixing with water can also be used as a foliar spray for small trees in the latter half of the growing season, but is less selective than fosamine.

Results have been mixed using the above treatments. Experiment on a small area to determine if treatment works or if sprouting occurs.

Multiflora Rose
Rosa multiflora

Multiflora rose has a spreading growth form with wide, arching canes.

A member of the Rose Family, multiflora (many flowered) rose is a dense, spreading shrub introduced from Japan and Korea for ornamental purposes in the mid to late 1800s. It was promoted in the 1930s to curb soil erosion, in the 1950s as a living fence to control livestock and create snow barriers along highways, and in the 1960s as a food source and cover for wildlife. It is also used as the rootstock for some grafted ornamental roses.

Because multiflora rose forms dense, impenetrable thickets that exclude other vegetation, it has become a serious problem in many states, especially in the southern region of the upper Midwest. It is particularly destructive to pastureland because it replaces many of the forage plants needed by livestock.

By law, multiflora rose is considered a nuisance weed in Wisconsin and may not be sold or propagated. It is classified as noxious in several other states, including Iowa, Indiana, and Missouri.

Habitat: Old fields, pastures, prairies, roadsides, open woods; can tolerate a wide range of soil and environmental conditions; thrives in full and partial sun with well-drained soils; is not found in standing water or extremely dry habitats; cannot tolerate winter temperatures below −28°F, limiting its northern range.

Height and Growth Form: Can reach 10–15′ in height and 9–13′ wide; root crowns of older plants can be 8″ or more in diameter; typically more spreading than erect.

Leaves: Pinnately compound; alternate; usually consist of 5 to 11 small (0.5–1″ long) oval leaflets with toothed margins; leaflets are nearly smooth on upper surfaces and paler with short hairs on undersides; petioles have feathery stipules.

Flowers: White; 0.5–1″-wide; 5 petals; form a panicle; bloom in May or June.

Clusters of white flowers bloom in May or June.

Fruit: Red; small (0.25″ across); hard; nearly round; occur in clusters; form in August; remain on plant throughout winter; eaten by many birds and mammals, which disperse the seeds.

Seeds: Remain viable in the soil for 10 to 20 years; individual plants can produce 500,000 seeds per year.

Stems: Wide, arching canes; covered with hard, curved thorns; branched; woody; up to 13′ long; typically grow upward for about 6′, then bend to the ground; capable of rooting when in contact with the soil.

Reproduction: By seed, horizontal stems (stolons) that can root at the nodes, and by shoots that root at their tips.

Control:

Manual or Mechanical Control: After removing their thorny tops, small plants can be dug with a shovel

Fruits are nearly round and occur in clusters. Leaves have small leaflets with toothed margins; petioles have feathery stipules.

or pulled with the help of a leverage tool (see chapter 2, "Weeding Tools"). All roots must be removed since new plants can develop from root fragments. Pulling plants out with a chain and tractor and bulldozing are also effective but cause severe soil disturbance.

Mowing with heavy equipment or cutting has also proven effective. Cutting should be repeated three to six times during the growing season for at least two to four years. Be aware that the thorns of multiflora rose may puncture vehicle tires.

Controlled burns in early spring may be helpful for infestations in fire-adapted plant communities. Follow-up monitoring is necessary to prevent new plants from establishing from the seed bank or from remaining root fragments.

Chemical Control: Cut-surface treatment using a 20% a.i. solution of glyphosate is effective on multiflora rose.

Basal bark treatment using a solution of 12.5% a.i. triclopyr formulated for use with penetrating oil has also been effective in controlling this species.

For foliar treatments, a 2% a.i. solution of fosamine in water can be applied from July to September. The foliage must be well covered but not dripping. Dieback will not be apparent until the following summer. Fosamine is often the preferred foliar spray treatment for multiflora rose since it is a bud inhibitor that selectively controls woody species.

A broadleaf-specific solution of metsulfuron-methyl plus a surfactant can also be used as a foliar treatment.

For a more selective method of foliar herbicide application, mow the site first and spot treat resprouts with a broadleaf specific herbicide such as triclopyr formulated for use with water.

Foliar applications of 1% a.i. glyphosate can be applied to flowering or budding plants. Glyphosate is nonselective, so care must be taken to avoid nontarget species. Do not use glyphosate as a foliar spray in high-quality natural areas.

Nonselective herbicide in granules or pellets can be thrown at the base of the rose without getting too close to the thorns. While effective, some of these herbicides may sterilize the ground for several years.

Biological Control: Multiflora rose is vulnerable to defoliation by Japanese beetles, another introduced species that has proven very troublesome for the turfgrass and ornamental landscape industry. Multiflora rose contains a substance, known as geraniol, that attracts Japanese beetles. Multiflora rose also suffers from rose rosette disease, which is a viruslike disease, the vector for which is a native eriophyid mite. The disease causes infected rose plants to turn a deep red color, sprout witches' broom-like growth, and produce more thorns than usual. Infected plants are usually dead in less than two years. Introducing rose rosette disease to roses generally involves grafting infected stems onto the target plants. If ornamental roses are growing downwind from the infected multiflora roses, they may also become infected.

Vines

Oriental Bittersweet
Celastrus orbiculatus

Oriental bittersweet, a rapidly spreading woody vine, is best known for the colorful orange berries that it produces in fall. Native to eastern China, Korea, and Japan, it is often planted as an ornamental vine that twines along fences or for use in dried craft arrangements. Birds eat its berries and spread seeds to new locations. Oriental bittersweet is often found invading forests and eventually forms dense stands by shading out other vegetation. After covering the ground and eliminating spring ephemerals, it climbs over nearby trees and shrubs. Trees can be girdled by the vine's woody stem. They can also become susceptible to wind damage as their crowns are weighted down. Oriental bittersweet also hybridizes with American bittersweet making it a genetic threat to this native species.

Round or oval leaves help differentiate Oriental bittersweet from the pointed-leaved American bittersweet.

Other common names: Asian or round-leaved bittersweet.

Habitat: Grasslands, open woods, woodland edges, undisturbed forests, roadsides, fencerows; shade-tolerant.

Leaves: Simple; alternate; rounded; glossy; finely toothed; 2–5″ long.

Flowers: Small; greenish yellow; 5-petaled; clustered in leaf axils.

Fruits: Green during summer, become yellow to orange in fall; rounded; 1 to 3 clustered in the leaf axils; comprised of 3-sectioned capsules with orange-red seeds.

Stems/vines: Can climb 60′ high in trees and reach 4″ in diameter; often have noticeable lenticels; older plants become more shrublike with spreading, trailing stems.

Reproduction: By prolific seed production and spreading underground roots that form new stems.

Similar species: American or climbing bittersweet (*Celastrus scandens*) is very similar in appearance to Oriental bittersweet. American bittersweet, however, has elliptical rather than rounded leaves, and its flowers and fruits occur at the ends of stems rather than along the stems in leaf axils.

Control:

Manual or Mechanical Control: Pull or dig up small infestations and remove from the area.

Chemical Control: For more established invasions, triclopyr formulated for use with penetrating oil or a strong solution of glyphosate (20% a.i. suggested) can be applied to cut stems in the fall after native plants have gone dormant or in early spring before the emergence of spring ephemerals. On severely disturbed sites, a mixture of 2,4-D and triclopyr can be carefully sprayed on foliage immediately after the first hard frost in the fall. Monitor for new infestations and remove them immediately.

FORBS

Garlic Mustard
Alliaria petiolata

Rosette-leaved colonies of garlic mustard stay green during the winter and are sometimes confused with creeping Charlie (Glechoma hederacea).

Garlic mustard is a rapidly spreading herb that is a major threat to the survival of native woodland vegetation and the wildlife that depends on it. Brought to the United States by early settlers for use in cooking and as medicine, this cool-season biennial forms dense colonies and spreads readily into both high quality forests and disturbed habitats.

Beginning growth early in the spring, garlic mustard can shade or crowd out native wildflowers and tree seedlings before they have a chance to grow. It can totally dominate a forest floor within five to seven years of its introduction.

Garlic Mustard

This pernicious weed [garlic mustard] is becoming a *very* serious matter that the public *must be made aware of.* Botanically speaking, it has all the makings of a tragedy for the lovely herbaceous flora of our deciduous forests. (Hugh Iltis, Director Emeritus, University of Wisconsin Herbarium, 2000)

Garlic mustard . . . is such a serious and menacing weed that the University of Wisconsin Arboretum in Madison has quarantined areas infested with garlic mustard in an attempt to squelch its spread. . . . The arboretum is also committing thousands of dollars and thousands of hours of volunteer and staff labor to try to control this weed that showed up just three years ago. . . . Garlic mustard has the ability to quickly spread across high quality woodland, seriously threatening native plant species and the wildlife that depend upon them. (Laurie Weiss, University of Wisconsin Extension–Milwaukee County, Commercial Horticulture Agent, *Community Newspapers, Inc.,* August 17, 1995)

The potential losses that are being caused by this plant [garlic mustard] are preventable and reversible—but only if we act now to get the information out to landowners, land managers, and park visitors. We need everyone's cooperation to prevent this plant from moving into currently uninfested woodland and to contain it where it has already spread. (George Meyer, secretary, Wisconsin Dept. of Natural Resources, 2000)

Dense populations rapidly displace native woodland plants.

Habitat: Most often found in shaded areas, also in full sun; upland and flood-plain forests, along trails, at the base of large trees; savannas, yards, roadsides, stormwater entry points; cannot tolerate acidic soils.

Odor: Leaves and stems emit the distinct odor of garlic when crushed, particularly in spring and early summer.

First-year plant: A rosette; has one to several, scallop-edged, dark green leaves; rises 1–6″ above the ground; growth begins in early April; rosette remains green through the following winter, making it possible to check for the presence of garlic mustard throughout the year.

First-year, rosette-shaped leaves have scalloped edges.

Garlic mustard in late June.

Second-year plant: A stalk emerges from the rapidly growing basal rosette in April or May of the second year, flowers are produced, seed pods begin to form shortly after flowers open, seed-drop begins in July, and the plant dies.

Height: Ranges from 1–48″ tall when in flower, generally between 10–30″. (No native, white woodland wildflower in this region is this tall in May.)

Leaves: Simple; rounded to heart-shaped at the base, becoming more triangular higher on the flower stalk; have large teeth; alternate on stems; can reach 2–3″ across; leaf stalks are longer on leaves toward the base of the plant. Smell like garlic when crushed.

Flowers: Numerous, white, small; have four separate 0.25″-wide petals; usually in clusters at the tops of stalks; sometimes in leaf axils; normally bloom from late April or early May through early June.

Fruit: Slender capsules; 1–2.5″ long; begin to form on lengthening stems shortly after flowering begins.

Seeds: Sole means of reproduction; produced by fruits in a single row; viable within days of initial flowering; oblong; black; have ridged coating; hundreds per plant; disseminated in July and August when plant dies; dispersed on the fur or feet of mammals such as deer, horses, and squirrels, by flowing water, and by human activities; can remain viable for at least seven years; two periods of germination, one in midspring and another, possibly longer period, in late summer.

Stems: Generally one or two, can be more, especially if tops have been cut or pulled off. Smell like garlic when crushed.

Taproot: Slender; white; *s*-shaped at the top; has multiple adventitious buds that can develop into flowering stalks if only the top of the plant is removed.

Second year: stalk develops; leaves become triangular with toothy margins; small, four-petalled flowers bloom in May.

Seed capsules develop soon after flowering begins.

Plants die when seeds mature; note rosette-leaved, first year plants below dead foliage.

Seeds become viable quickly and begin to disperse in July.

Control: Always clean shoes, pant cuffs, camping equipment, bike tires, and other items carefully before leaving garlic mustard infested areas to prevent the spread of seeds. Because garlic mustard seeds remain viable for at least seven years, infested areas need to be monitored after initial control efforts begin. It is crucial that uninfested woodlands are also monitored to spot new invasions early. Removing a few plants before they go to seed is much easier than removing the large number of plants that will result if any plants are allowed to disperse their seeds.

Manual or Mechanical Control: There are three basic methods of manual or mechanical control for garlic mustard: pulling, cutting, and burning.

Pulling: First-year plants may be pulled any time the soil is not frozen. All plants—both first- and second-year—should be pulled from new, small infestations. Because dense patches of new seedlings self-select, control efforts usually emphasize removing second-year plants before they disperse their seeds. For larger patches, determine the outer edges of an infestation and remove plants working from least-infested areas to most-infested areas, or determine a priority site to protect and remove garlic mustard plants moving out from that area. Pull slowly from the base of plants to keep them from breaking off. Take care to remove the entire root to prevent plants from resprouting.

Plants pulled before flowers open should be scattered about to help them dry. Plants left in piles or in moist conditions may go on to flower and produce seed. Once flowering has begun, pulled plants should be bagged (see Disposal). Pulled plants can be placed in a row with all flower or seed heads going in one direction. Flower or seed heads can then be burned off with a propane torch.

It is usually better to weed an area thoroughly rather than try to cover too much territory. It is very important to revisit weeded areas at regular intervals in the spring to find missed plants.

Cutting: Cutting garlic mustard plants is faster than pulling and avoids disturbing the soil, which may bring more seeds to the surface and encourage germination. A string trimmer, scythe, or power brush cutter may be used. However, sensitive wildflowers and tree or shrub seedlings may be cut and possibly killed in the process. Therefore, before cutting, evaluate the site to determine how best to proceed.

Studies show that cutting produces mixed results. Many researchers recommend cutting garlic mustard as close to the soil surface as possible just after flower stalks have elongated but before flowers have opened. Cutting a couple inches above ground level is not as effective, and cutting before stem elongation may lead to resprouting with multiple stems. This procedure, according to some studies, has resulted in killing up to 99 percent of the plants. Other researchers say they have had significant resprouting when using this technique. Results may depend upon timing and field conditions.

If flowering has begun, cut plants must be bagged and properly disposed of (see Disposal). Because garlic mustard plants in one location are often at various stages of development, it may be difficult to cut all plants at the optimum time. Cutting may also expose small, previously hidden plants to more light, allowing them to grow more vigorously and mature after cutting takes place. Infested areas should therefore be revisited after initial cutting to remove any flowering plants.

If seedpods are well developed and stems have become tough and fibrous, plants can be cut at ground level before seed drop begins, and then bagged. Resprouting is unlikely at this late stage of development, but sites should be rechecked.

Disposal: There are several methods for disposing pulled or cut garlic mustard. Land managers should evaluate their particular site to determine the method most appropriate for their situation.

- If garlic mustard must be controlled over a large expanse of land and if infestations are severe in some areas, pulled or cut plants from the least affected areas can be deposited in the center of a severely affected area until control efforts can take place there.

- Pulled or cut plants can be scattered on a lawn near the infested site. Be careful not to spread them so deeply as to smoother the grass. Using a mulching lawn mower, chop plants into small pieces and continue to mow the area at regular intervals during the growing season. Mowers should not be directed to discharge into a natural or wooded area and should be cleaned before moving on to a new site. Initial results using this method of disposal have been good. Do not spread plants out on lawns adjacent to natural areas that are free of garlic mustard.

- Plants can be bagged and later buried deeply in an area that will not be disturbed. Landfilling plant material may not be legal in your area. If your local waste hauler refuses to pick up bags of garlic mustard, call your natural resources department and ask whether there are alternative means of disposal. Composting is not recommended as the seeds may survive and be transported to new sites along with the compost.

- Dried plants can be placed in large, paper composting bags (available at many hardware and garden stores) and burned. To promote drying, the bags should be left open and protected from rain. Small holes can be poked in the sides of the bags to speed the process. Due to cost of the bags, this method works best on small sites and in situations where smaller plants have been pulled or cut early in the growing season.

Burning: For large infestations, fall or early spring burning in oak woodlands or fire-adapted plant communities may help control garlic mustard. First-year plants, which remain green over winter, are killed by fire if it is hot enough to remove all leaf litter. Three to five consecutive years of burning are recommended because burning will release garlic mustard seeds in the seedbank. Any surviving garlic mustard plants should be pulled or killed with a propane torch or herbicide after a burn. Fall burning is usually more effective since leaf litter is more likely to provide adequate fuel. Spring burning must be done early enough to minimize injury to spring wildflowers.

Propane torches with elongated wands have been used successfully on garlic mustard seedlings during periods of wet weather in locations where regular burning is not allowed. [R. Retko] This method of control may be helpful in urban areas where controlled burns are prohibited or when herbicides cannot be used.

Chemical Control: Severe infestations can be controlled by applying a 1 to 2% a.i. solution of glyphosate to the foliage in early spring. At this time of the year, most native plants are dormant, while garlic mustard is green and vulnerable. Ideally, another application can be made to new seedlings soon after the spring germination period but great care must be taken to avoid native plants that may have emerged. A similar application of glyphosate can be made in the fall after a hard frost, but only if temperatures remain above 50°F for five consecutive days.

Leaf blowers can help expose first-year plants for herbicide applications. After glyphosate has been applied and die-off has occurred, smaller, previously hidden plants may now receive more light and begin growing more vigorously. Since they may not be noticed in spring until after native plants have emerged, pulling or cutting these plants may be necessary after herbicide use.

Broadleaf specific herbicides, such as 1% a.i. triclopyr formulated for use with water or 2,4-D amine, can be used near lawns or in grassy areas. Be sure chemicals have had adequate time to break down before allowing anyone in the area, especially children and pets.

Pelargonic acid, a fatty acid based, nonselective contact herbicide, has been used with some success on first-year plants up until about mid-June [K. Solis]. More than one treatment may be necessary for maximum control. Avoid contact with nontarget species.

A broadleaf specific solution of metsulfuron-methyl plus a surfactant can also be used as a foliar treatment. This mixture will have preemergent control of germinating seedlings.

Preliminary testing of glufosinate-ammonium on rosettes in fall or early spring has been encouraging so far [K. Solis]. This product rapidly degrades and does not have residual soil activity.

Early experiments with corn gluten applied to the soil where seeds have dispersed have shown some success as a preemergent alternative to herbicides for killing seedlings.

Biological Control: Researchers are studying ways to effectively control garlic mustard with biological agents. Several insects are being tested and may be available for limited releases in the United States in 2005 or 2006. For the latest information, contact the Cornell University, Biological Control of Non-Indigenous Plants Species Program, Ithaca, NY (607) 255-5314, or check their website, www.invasiveplants.net.

Spotted Knapweed
Centaurea maculosa syn. *C. biebersteinii*

Spotted knapweed has taken over this pasture, greatly reducing its forage potential.

Spotted knapweed is a short-lived, perennial herb of the Composite family. Each plant typically lives from three to nine years. Native to Eurasia, spotted knapweed was probably accidentally introduced in the 1890s as a contaminant in alfalfa or hay seed. It has become a serious problem on the rangelands of the northern plains, Rocky Mountain states, and Canadian provinces. In recent years, spotted knapweed has spread rapidly in the upper Midwest, invading even relatively undisturbed natural areas. It releases a toxin into the soil that hinders or prevents the growth of neighboring species. This promotes its domination, reduces plant diversity, and limits forage and crop production. Livestock will eat knapweed only when other vegetation is unavailable. As spotted knapweed populations rise and other plant species are excluded, surface runoff and sedimentation often increase. The water-holding capacity of the soil decreases as knapweed taproots replace the interconnected network of native plant root systems. Spotted knapweed is considered noxious in fifteen states, including Minnesota.

Habitat: Heavily disturbed sites, roadsides, agricultural field margins, undisturbed dry prairies, oak and pine barrens, rangeland, lake dunes, sandy ridges; most common in sunny habitats with well-drained or gravelly soils.

Spotted knapweed

In Montana, knapweed infestations result in an estimated direct annual impact of $14,000,000 with total (direct plus secondary) impacts of about $42,000,000 per year, which could support over 500 jobs in the state's economy. (S. Hirsch and J. Leitch, *The Impact of Knapweed on Montana's Economy,* Report # 355. North Dakota State University, Fargo, ND, 1996)

Studies in Montana show that spotted knapweed invasions reduced available winter forage for elk between 50 and 90 percent. (Dawn Bebunah and Jay Carpenter, *Proceedings of the 1989 Spotted Knapweed Symposium.* Montana State University, Bozeman, MT)

In a study area in Montana, runoff and sediment yield were 56 percent and 192 percent higher, respectively, for areas dominated by spotted knapweed than for native bunchgrass vegetation types. (J. R. Lacey, C. B. Marlow, J. R. Lane, "Influence of spotted knapweed on surface runoff and sediment yield," *Weed Technology Journal* 3:627–31)

Spotted knapweed rosette; leaves deeply indented.

Height: 2′ on upland sites to 4′ on wetter sites.

Leaves: Pale or grayish green; rough with fine hairs.

> *First-year rosettes:* Lower leaves are up to 6″ long; compound with deeply divided leaflets or with narrow, irregular lobes; leaf stems are long; plants live one to four years at this stage.
>
> *Flowering stems:* Leaves on the lower half of plant resemble those of the rosette; upper leaves are simple, alternate, linear with entire margins, 1–3″ in length, and become smaller near the top of plant.

Flowers: Numerous; thistlelike; pinkish or lavender to purple, fade to white during seed development; 0.75″ in diameter; occur singly on tips of branching stems; flower heads have stiff bracts with fine, vertical streaks and black, bristly hairs at the end; plants bloom late June through September; individual flowers bloom for two to six days each.

Seeds: Sole means of reproduction; about 1,000 per plant; brownish; under 0.25″ in length; notched on one side of base; germinate throughout growing season; viable up to nine years; wind dispersal begins about twenty days after individual flowers finish blooming; weight usually prevents dispersal more than 3′ from parent plant.

Seedlings: Emerge in fall; become rosettes that resume growth in spring.

Stems: Slender; rough; erect; 1–6 (up to 20) per plant; branched; bolting begins in early June.

Taproot: Stout; elongated; some plants also produce fibrous lateral roots that extend just below the soil surface for several feet.

Reproduction: Primarily by seed production; to a lesser extent by sprouting from lateral roots.

Similar species: Related species, Russian knapweed (*C. repens*), white-flowered or diffuse knapweed (*C. diffusa*), yellow starthistle (*C. solstitialis*), and bachelor's

Thistlelike flowers are pinkish purple; upper leaves narrow.

button (*C. cyanus*), also known as cornflower. These species are also invasive, but none of them, with the exception of bachelor's buttons, a popular garden plant often used in wildflower seed mixes, and Russian knapweed are as yet common in the upper Midwest. Bachelor's button has bright blue flowers that are generally larger than those of spotted knapweed.

Control: Preventing spotted knapweed from being introduced to an area is the most important element of control. Spotted knapweed seeds are often spread by rodents, livestock, contaminated hay, or vehicle undercarriages, mowers, and other equipment. Use caution when using hay from ditches of primary roadways or when buying hay from knapweed-infested areas. In order to deplete the knapweed seedbank, it may be necessary to use annual control measures on infestations for several years.

Manual or Mechanical Control: Early detection and removal of colonizing plants is vital for easy control or elimination of spotted knapweed. Outlying plants should be controlled before main populations. Small infestations can be removed by digging or pulling when the soil is moist.

CAUTION: Some individuals experience skin reactions following spotted knapweed exposure. Wear proper clothing (gloves, long pants and sleeves, etc.) when handling this species.

The entire root should be removed to prevent resprouting. Pulling is easiest on sandy soils after stems emerge and are big enough to grasp. Flowering plants should be bagged, removed from the site, and properly disposed of to make sure that seeds do not mature. Attempting to pull plants in the rosette stage often results in breaking the plant off at the crown, unless roots are loosened with a hand trowel or small shovel. A disadvantage of pulling is that it may bring more seeds to the soil surface and result in more germination. Revisiting the site for several years to eliminate new plants is essential.

Annual burns have reduced spotted knapweed populations from 5 to 90 percent. Reductions seem to be related to the intensity of the burn, which is dependent on the amount of grasses and sedges present. Burns that remove nearly all the duff are most effective at killing knapweed roots and normally succeed only in newly infested areas. Before using an intense burn, consider its impact on native plants in the area. If burns have effectively reduced most of the population, remove any remaining plants by pulling and digging. Reseed burned areas with native species, if they fail to reemerge.

Mowing at the start of flowering can help limit seed production. Mowing later may spread viable seeds.

Plowing may kill the current knapweed population and other existing vegetation, but new knapweed plants will undoubtedly emerge from the seedbank.

Grazing by sheep and goats may decrease spotted knapweed seed production for that year but will likely damage other vegetation.

Chemical Control: For chemical applications, include areas 10–15′ beyond knapweed populations to control roots and seedlings. Follow-up treatments are vital for continued control.

Clopyralid is effective on spotted knapweed and is more selective than some herbicides. Since it primarily affects legumes, composites, and plants in the Buckwheat family, clopyralid may be used on grazing land but should not be used on cropland areas. Do not apply in areas where soils have rapid permeability (such as loamy sand to sand) or where the water table is shallow, since groundwater contamination may occur.

Spotted knapweed rosettes can be treated with 2,4-D water soluble amine formulation in the fall or early spring. Use of glyphosate at these times has had mixed results. Applications should be made before stem elongation. Although 2,4-D may be less expensive than clopyralid, it will effect a much wider range of species and may not be as effective for control.

Biological Controls: Several insects—seed-head flies (*Urophora affinis* and *U. quadrifasciata*) and a moth (*Agapet zoegana*) whose larvae feeds on the roots of spotted knapweed—have been released in the upper Midwest and show promise in controlling spotted knapweed. Other promising biocontrol insects have been released farther west. Contact your local natural resources department for additional information or check the website paipm.cas.psu.edu/BenefInsects/beneficials_plantfeedrs.htm.

Canada Thistle
Cirsium arvense

A Canada thistle colony going to seed.

Canada thistle

Canada thistle is declared a "noxious weed" throughout the U.S. and has long been recognized as a major agricultural pest, costing tens of millions of dollars in direct crop losses annually and additional millions for control. Only recently have the harmful impacts of Canada thistle to native species and natural ecosystems received notable attention. (Plant Conservation Alliance, Alien Plant Working Group. www.nps.gov/plants/alien/fact/ciar1.htm)

Canada thistle, native to Eurasia, is an aggressive, dioecious perennial forb. Introduced to Canada in the seventeenth century in contaminated crop seed, this species forms huge monocultures by spreading rhizomes and massive seed production. It is a major agricultural pest but generally does not pose a serious threat to high-quality prairies. Canada thistle can, however, greatly reduce species diversity in old fields, disturbed natural areas, or areas under restoration. It is important to control this species prior to initiating restoration work.

Canada thistle is classified by law as a noxious weed in forty-three states, including Iowa, Illinois, Minnesota, Missouri, and Wisconsin. It should not be allowed to set seed.

Leaves, somewhat lobed, have crinkled edges with numerous spines.

Lavender flowers bloom July to September.

Habitat: Disturbed open areas, roadsides, agricultural fields, prairies, savannas, clay to gravelly soils; sometimes found in wetlands if dry periods occur.

Leaves: Simple; alternate; lance-shaped; tapering; irregularly lobed; crinkly with spiny, toothed margins; stalkless; green on both sides; smooth on top, sometimes woolly beneath.

Flowers: Numerous; rose-purple to lavender, sometimes white; 0.375–0.625″ in diameter; typically clustered; bracts have spineless tips; bloom June through September; fragrant.

Note: The native swamp thistle (*C. miticum*), which is necessary for the survival of the swamp metalmark butterfly, has a flower head similar to that of Canada thistle. Unlike Canada thistle's lavender flowers, however, swamp thistle flowers are pink and sticky behind the flower head.

Seeds: Small; light brown; slightly tapered; 0.18″ long; tuft of tan hair loosely attached to the tip enables wind dispersal but easily breaks off causing most seeds to fall near parent plant; 1,500–5,000 seeds produced by one plant; capable of germinating eight to ten days after flowers open, even if plants were cut in bloom; most germinate within one year, but some remain viable in the soil for twenty years making eradication difficult; can be spread by mowing equipment once flowering begins; usually dispersed by wind; grows in clonal patches of all female or all male plants; only female patches produce seeds.

Stems: Upright; 2–5′; slender; ridged; branch only at the top; slightly hairy when young, becoming more hairy with age.

Root system: The main means of reproduction; the fibrous taproot may extend 6′ deep in loose, well-tilled soil; horizontal roots stemming from the taproot of a single plant are usually within 1′ of the soil surface and typically spread 3–6′ in one year but are capable of spreading 10–18′; aerial shoots are sent up in 2–6″ intervals and generally produce basal leaves the first year and flowering stems the next year; small root fragments are capable of producing new shoots within five days.

Control: Encouraging the development of healthy, dense native prairie vegetation can help prevent and reduce Canada thistle. A buffer zone free of Canada thistle should be maintained near important natural areas to prevent invasion. Plants within wind dispersal range of a site about to undergo restoration should be controlled prior to commencing work.

Manual or Mechanical Control: In a high-quality natural area, repeated mowing or selective cutting close to the ground can reduce an infestation within three or four years. Use a scythe or other sharp tool for selective cutting. The ideal time to cut or mow Canada thistle is prior to the flower buds opening because the plant's food reserves are at their lowest point and seeds have yet to form. Cutting and mowing should be done at least three times during the growing season (June, August, and September). Mowing after flowering will only spread the seed. Mowers should be cleaned if used in areas with thistle in seed. Plants cut eight or more days after flowers have opened should be collected and removed from the site to prevent seed development from progressing. Cutting new shoots every twenty to thirty days for two or three seasons can reportedly eradicate individual populations.

Early spring burns tend to increase sprouting and reproduction. Late spring burns may discourage established Canada thistle plants but may also encourage further seed germination. Burns may, therefore, need to be conducted for three consecutive years.

Chemical Control: Control of this species with herbicides in high-quality natural areas is not recommended because herbicides can cause more damage to native vegetation than the damage caused by Canada thistle. In other areas, clopyralid or metsulfuron-methyl have been used effectively as foliar sprays. Researchers report good results from using a 1–2% a.i. solution of glyphosate during the early bolting stage when plants are 6–10″ tall, during the bud to flowering stage, or when applied to rosettes in the fall. Remember that glyphosate is nonselective and will kill or harm any green plant that it contacts.

Spot applications to control individual plants or stems are most effective with a wick applicator or hand sprayer using either clopyralid or glyphosate.

Crown Vetch
Coronilla varia

Crown vetch, an herbaceous perennial, is a member of the Legume family. It is native to Europe, southeast Asia, and northern Africa. Crown vetch is frequently planted along roadsides and waterways for erosion control, and sometimes as a green fertilizer crop. It forms large, dense mounds of vegetation that climb over and shade out other species. Because it seeds easily and spreads rapidly by creeping roots, crown vetch easily escapes from cultivated areas making it a serious management threat to native plant communities. Although crown vetch may be acceptable for limited use in urban settings where its spread can be contained, there are alternatives for erosion control. Encourage your local highway department to stop planting crown vetch and consider less invasive species for roadside vegetation.

Crown vetch

Roadside rights-of-way account for more than 10 million acres of land in the United States. This land requires care that assures water quality, improves erosion control, protects wildlife habitat, reduces mowing and spraying, enhances natural beauty, controls noxious weeds, and protects our natural heritage . . . all objectives of integrated vegetation management.
(Bonnie Harper Lore, Federal Highway Administration, 2001)

Once used widely along highways, many highway departments no longer add crown vetch to their seed mixes.

Habitat: Prefers full sun; tolerates partial shade and many soil types; spreads from plantings into remnant prairies, agricultural fields, pastures, roadsides, waste places, woodland edges, and along streambanks.

Leaves: Pinnately compound; alternate; 2–6″ long leaflets; no leaf stalks; consist of 15–25 pairs of oval, 0.75″-long leaflets.

Flowers: Pinkish lavender and white; pealike; arranged in rounded, 1″-wide clusters on long, extended stalks that grow from leaf axils; bloom May or June through August.

Fruit: Long, narrow seedpods; contain a few to several slender, brown seeds that can remain viable for more than fifteen years.

Stems: 2–6′ long; trail along the ground as they grow; branched; hairless; plant cover rises 6–12″ above the ground.

Root system: Rhizomes, up to 10′ long, produce roots on the lower surface and shoots on the upper surface; roots are not fibrous, which limits the ability to control erosion.

Dormant appearance: Large, brown, earth-hugging patches.

Reproduction: By seeds and creeping root system.

Control:

Manual or Mechanical Control: In plant communities that tolerate fire, such as prairies, controlled burning in late spring can be effective. Burns may need to be repeated for several years for adequate control. Repeated hand-pulling can be effective in smaller infestations.

Late spring mowing for several successive years can also help control this species. Another technique is to mow twice every year, in June and late August, to cor-

respond with each leaf-out period. Small patches can be controlled by covering them with black plastic.

Chemical Control: Spring burning will help eliminate the previous year's accumulation of old crown vetch stems and leaves and expose new growth. This will help facilitate adequate herbicide coverage on new leaves. Follow-up applications may be necessary the following fall or early spring to combat regeneration from seeds or rhizomes. Avoid herbicide contact with nontarget plants.

Clopyralid, which is often used to treat populations of crown vetch, is more selective than the chemicals discussed below, and will likely be more effective.

2,4-D amine or a solution of 2% a.i. triclopyr formulated for use in water are broadleaf specific and can be applied to crown vetch foliage in early spring when it is actively growing. Use a hand sprayer for spot applications.

A 1 or 2% a.i. solution of glyphosate, which is nonselective, can also be applied to foliage in early spring.

Flowers are pinkish lavender to white; compound leaves have paired leaflets; seedpods are long and narrow.

Teasels

Common Teasel: *Dipsacus sylvestris*
Cut-leaved Teasel: *Dipsacus laciniatus*

Cut-leaved teasel rosette.

Teasels can quickly take over a disturbed area. The dried plants are often used in craft arrangements.

Common teasel has purple flowers.

Teasels are aggressive, monocarpic perennials. Introduced to North America from Eurasia and northern Africa in the 1700s for combing wool, they are now popular in dried flower arrangements and horticultural plantings. Teasels can form extensive monocultures and have become a severe threat to some Midwestern natural areas. Although common teasel is more widely distributed in the United States, cut-leaved teasel is more frequently observed in the upper Midwest. Its range has been expanding rapidly in recent years. Common and cut-leaved teasel may also hybridize.

Habitat: Open, sunny areas; wet to dry soils; seem to prefer mesic habitats; most commonly seen along roadsides and in heavily disturbed areas; also found in high-quality prairies, savannas, and sedge meadows.

Leaves: Simple; opposite; flowering plants have large, oblong leaves that attach to stems without petioles; prickly, especially on lower side of midrib; form cups that hold water where they attach to stems. Somewhat oval on young rosettes; become large, oblong, and quite hairy on older rosettes.

Cut-leaved teasel has white flowers.

>   *Common teasel:* Leaves are not divided.
>   *Cut-leaved teasel:* Leaves have feathering deep lobes; broader than those of common teasel.

Flowers: Small; tubular; packed in dense, oval-shaped, spiny heads at the ends of stems.

>   *Common teasel:* Generally has pink or purple flowers; blooms June through October.
>   *Cut-leaved teasel:* Usually has white flowers; blooms July through September.

Bracts: Curve upward from below flower heads; stiff and spiny.

Common teasel: Bracts are longer than the flowering head.
Cut-leaved teasel: Bracts are usually shorter than the flower head.

Seeds: One plant can produce more than 2,000 seeds; 30–80 percent may germinate; remain viable for at least two years; typically do not disperse far from parent plant; mowing equipment, flooding and careless disposal of dried flower arrangements can increase spread.

Stems: Upright; prickly; angled; may reach 6–7′ in height; dead stems remain upright through winter.

Taproot: Becomes large during the rosette stage; grows to over 2′ in length and 1″ in diameter at the crown.

Life cycle: Grows as a basal rosette for at least one year; a tall, flowering stalk then emerges; plant dies after flowering and producing seed.

Control: Do not buy dried flower arrangements containing teasel. When necessary to dispose of seed heads, burn or bury them deeply in an area that will not be disturbed. Control methods will likely need to be repeated for several years due to viable seeds remaining in the soil.

Manual or Mechanical Control: For small infestations, teasel rosettes can be dug using a dandelion digger. Remove as much of the root as possible to prevent resprouting. Cutting through the taproot below the soil surface with a sharp shovel can be helpful, but the area should be checked later for resprouts. An alternative method of control involves cutting the stalks just before flowering. Plants are unlikely to reflower, but will die at the end of the growing season. If flowers have opened, cut stalks should be removed from the natural area since seeds can mature on stems after cutting. Dispose of properly. Avoid cutting before full bud stage since plants are likely to send up new flowering stalks.

If teasel has not become dense and adequate amounts of dried grasses or sedges are present for fuel, late spring burns can be helpful. Burning should be used with other control methods for maximum impact. Burning a site in the spring so that rosettes can easily be seen on the blackened soil can help in manual control.

Chemical Control: Teasel rosettes remain green late into fall and begin growing early in the spring when most native plants are dormant. Herbicide application at these times reduces the risk of harming nontarget species. Temperatures should be 50° F or warmer for maximum effect.

2,4-D amine is selective and economical and has been used successfully to treat teasel when applied before flower stalks emerge.

1.5–2% a.i. glyphosate or 2% a.i. triclopyr formulated for use with water can also be carefully applied to foliage and stems before flower stalks emerge. Triclopyr and clopyralid are apparently more effective than glyphosate and have the added advantage of being broadleaf specific. Herbicides may be applied after the flower stalks emerge, but seed development remains a risk. Avoid herbicide contact with nontarget species.

Leafy spurge

Leafy spurge has infested over 1.2 million acres of rangeland in North Dakota alone, resulting in an estimated annual loss of nearly $75 million. (Rodney Lym, Calvin G. Massersmith, Richard Zollinger, "Leafy Spurge Identification and Control," NDSU Extension Service, Circulation # W-765, April 2000)

Annual economic impacts of leafy spurge infestations on grazing and wildlands in Montana, North Dakota, South Dakota, and Wyoming are about $129,000,000. (J. A. Leitch, F. Larry Leistritz, Dean A. Bangsund, "Economic Affect of leafy spurge in Upper Great Plains," Agriculture Economic Report, #316, North Dakota Agriculture Experiment Station)

[Because of leafy spurge], ranchers have had to abandon more than a million acres that could have supported 90,000 cattle. (Sharon Begley, "Aliens Invade America!" *Newsweek,* August 10, 1998)

In the mid-1980s, a 1,360-acre ranch in Klamath County, Oregon, was abandoned due to nonproductivity caused by leafy spurge. It was later purchased at an auction in 1988 for 17 percent of the value it would have sold for otherwise. (Randy Westbrooks, Invasive Plants—Changing the Landscape of America: Fact book. Federal Interagency Committee for the Management of Noxious and Exotic Weeds (FICMNEW), Washington, DC, 1998)

Leafy spurge is poisonous to cattle and causes severe eye irritation and possibly blindness in humans. (Bureau of Land Management, "The Spread of Invasive Weeds in Western Wildlands: A State of Biological Emergency," The Governor's Idaho Weed Summit, Boise, ID, May 19, 1998)

Leafy Spurge
Euphorbia esula

Leafy spurge, a member of the Spurge family, is legally classified as a noxious weed in nineteen states, including Iowa, Minnesota, and Wisconsin. Landowners in these states are required to eradicate it. Leafy spurge is a deep-rooted Eurasian perennial that can dramatically reduce the economic productivity and biological diversity of grasslands. It can reduce the productivity of grazing land by 50 to 75 percent because cattle and horses will not eat it. Leafy spurge currently inhabits about three million acres of rangeland in the United States with the heaviest concentrations in the northern Great Plains, particularly North Dakota. In just four states, the economic losses from leafy spurge exceed $100 million annually.

Leafy spurge outcompetes other vegetation by beginning growth early in the spring, shading-out competitors, and taking more than its share of moisture and nutrients from the soil. It also appears to produce chemicals that interfere with the growth of other plant species. In natural areas, leafy spurge destroys wildlife habitat by displacing native grasses and forbs in only a few years after its introduction.

Leafy spurge can take over expanses of former grasslands.

First recorded in the United States in 1827, leafy spurge may have arrived here accidentally as a contaminant in agricultural seed or intentionally as an ornamental plant.

Habitat: Open areas, but tolerates partial shade; rangeland, roadsides; prairies, savannas, gravel pits, damp to dry soils; most aggressive on semiarid sites.
Height: 6–36″.
Leaves: Simple; alternate; long; narrow; 0.25″-wide; bluish green; usually pointed and drooping with smooth margins; have petioles.

Flowers: Very small; green; petals fused into a cuplike structure; surrounded beneath by prominent bracts; bloom begins around mid-June, about two weeks after bracts appear, and continues to the end of July; some plants continue to bloom into fall; produce large amounts of pollen and nectar.

Bracts: The most colorful and conspicuous part of the plant; large; yellow-green; heart-shaped; petal-like; paired; seven to ten are borne in umbels at the tops of stems; open late May or early June.

Seeds: Oblong; gray-brown; smooth; produced in three-celled capsules (one seed per cell); capsules expel seeds passively in mid- to late-July, or explode open dispersing seeds up to 15′ away; can produce more than 200 seeds per plant; high germination rate with most seeds germinating within the first two years; typically germinate late May to early June but germination can occur throughout the growing season if there is adequate moisture; remain viable up to eight years; spread by wildlife, humans, and water.

Stems: Erect; branched at top; nonwoody; hairless; shoots emerge in late March; grow rapidly in May and June; can reach densities of up to 1,800 stems per square yard.

Sap: Milky white, sticky (latex); seeps from plant when any part is cut or torn; can help in identification, although it is present in all species of the Spurge family.

Roots: Woody; tough; brown; pink buds develop new roots or stems; form a dense network to 12″ deep; taproots reach up to 15′ deep; rhizomes can spread 35′ laterally; vegetative growth along rhizomes and root crowns is the primary method of expansion into undisturbed areas; root reserves enable plants to recover from most damage; very persistent; if foliage is destroyed, roots will generate new shoots.

Similar species: Cypress spurge (*E. cyparissias*), closely related to leafy spurge; grows 6–12″ high; introduced as an ornamental ground cover; especially used in old cemeteries; has been found invading dry grasslands; has more leaves than leafy spurge with each leaf being shorter (about 1″ long) and more narrow; will

Flowers bloom mid-June to late July; leaves are narrow; broken stems exude a sticky, milky sap.

Yellow-green flower bracts are conspicuous.

tolerate wetter, shadier conditions; some garden centers sell it; control techniques are similar to those for leafy spurge. Leafy spurge should not be confused with the flowering spurge (*E. corollata*), a native spurge with white flowers and more erect leaves.

A number of spurges hybridize with leafy spurge and are often collectively called "leafy spurge."

Control: No mechanical control methods have proven effective on leafy spurge. Hand-pulling, digging, and tilling succeed only if the entire root system is removed. The number of plants may actually increase if any root fragments remain in the soil.

Burning has not been very effective for leafy spurge control. In areas dominated by leafy spurge, there is little fuel (dry grasses, sedges, or leaf litter) and fires, therefore, do not burn hot enough to kill the plants. Studies indicate that while spring burning may decrease seed germination, it also stimulates leaf production.

Weekly mowing can reduce leafy spurge seed production, but does nothing to limit the spread of rhizomes.

Hay, grain, or gravel contaminated with spurge should not be moved. Cultivating equipment and vehicles used near spurge colonies should be checked before being moved to new areas to make sure that weed fragments have been cleared from equipment.

Since current methods used to control leafy spurge have failed to eradicate well-established populations, early detection is imperative.

Chemical Control: Until biological controls are easily available, experts advise land managers who find small infestations of leafy spurge to take immediate action to control it with chemical applications. Doing nothing will result in further spread, make future control efforts more expensive and less successful, and have a greater impact on native grasslands and rangelands.

Studies suggest that even when the best available herbicides are used, total elimination of a leafy spurge infestation is unlikely. Even when treatments are successful the first year, the population will likely recover within four years without additional treatment due to new shoots arising from persistent rhizomes and from seed germination.

Imazapic is a selective herbicide that has been effective on leafy spurge. It may harm cool-season grasses, but warm-season prairie grasses, such as big and little bluestem, Indiangrass and sideoats grama, tolerate its use. Imazapic may leach into groundwater on sandy soils, especially where the water table is shallow. Check with local cooperative extension soil experts before using this product. Imazapic should be applied in the fall before a killing frost.

Experiments in southern Wisconsin show that the combination of a controlled burn in the spring followed by a foliar application of fosamine after leafy spurge plants have resprouted and begun blooming (June/July) is an effective control measure. Using this method, control was achieved one year after the chemical application, although follow-up treatments were necessary for three to four years to kill plants germinating from the seedbank [S. Glass].

2,4-D amine plus glyphosate can be used to treat small patches of leafy spurge but is nonselective, requires repeated application, and has poor long-term control.

Biological Control: The United States Department of Agriculture has done extensive experiments involving six insects, including flea beetles of the genus *Aphthona*, to help control leafy spurge. Results thus far have been encouraging. It is important to match the various beetles with their appropriate habitat requirements. These beetles are available in most states with spurge infestations. Persons interested in releasing them may be able to collect extra beetles at sites where releases have been done several years previously. For more information, contact your state invasive plant control coordinator or the USDA's Team Leafy Spurge (www.team.ars.usda.gov/index.html).

Angora goats and Targhee sheep like to eat leafy spurge flowers, thereby reducing seed production. In native prairies, grazing is most helpful in the spring when the majority of native grasses are still dormant. Sheep and goats are generally less costly to use than chemical control measures but do not limit the spread of leafy spurge by rhizomes. Ranchers who raise only cattle should consider stocking goats or sheep if they have an extensive infestation of leafy spurge.

Bird's-Foot Trefoil

Lotus corniculatus

Bird's-foot trefoil, also known as deer vetch, is a perennial legume, native to Europe. It has been widely planted along roadsides for erosion control and is sold commercially as livestock forage and as a groundcover for landscaping. Many cultivars are available. Once established, bird's-foot trefoil creates tangled mats of dense growth that choke out other plants. Bird's-foot trefoil easily escapes cultivation, threatening the diversity of native plant communities. It tends to be more aggressive in the northern parts of the Midwest.

Habitat: Roadsides, fields, prairies, open disturbed areas, waste places; tolerates a wide variety of soil types and moisture conditions, including drought.
Height: Rises 6–24″ above the ground.
Leaves: Pinnately compound; alternate; cloverlike; without teeth; consisting of three oval leaflets that are about 0.5″ long and two smaller leafletlike stipules at the base of the leaf stalk.
Flowers: Bright yellow; sometimes streaked with red; pealike; 0.5″ long; borne in flat-topped clusters of three to six at ends of stems; bloom June to frost.

Bird's-foot trefoil is commonly used in roadside seed mixes.

Bird's-foot trefoil has yellow, pealike flowers.

Fruits: Clusters of 1″-long, slender brown to almost black seed pods; resemble a bird's foot; 10–20 seeds are ejected from each pod at maturity.

Stems: Sprawling to erect; several emerge from a single root crown; have many branches that become tangled and matted; can be smooth or hairy and up to 3′ long; become woody with age.

Root system: Taproot sometimes over 3′ long; secondary roots from rhizomes; above ground runners form a fibrous mat.

Reproduction: By seed, rhizomes, and above ground runners.

Control: See control methods for crown vetch controls, all of which are similar for bird's-foot trefoil.

Manual or Mechanical Control: Controlled burns increase seed germination and promote seedling establishment making bird's-foot trefoil a problem in native prairies. With small infestations, plants can sometimes be dug if soil is loose. Remove all root fragments. Frequent mowing for several years at a height of less than 2″ helps control this plant, but also sets back native species.

Chemical Control: Foliar herbicides containing clopyralid effectively control this species.

Purple Loosestrife
Lythrum salicaria

Wetlands can quickly become dominated by purple loosestrife.

Purple loosestrife, a wetland perennial from Europe, was introduced to the United States in the 1800s as a medicinal herb, nectar plant for honey bees, and garden plant. It also arrived here unintentionally as a contam-

---

Purple loosestrife

It is at once a beauty and a beast. In its seductive purple splendor, it spells death for everything in its way. . . . Purple loosestrife . . . has no natural predators in this country. It displaces native plant species (and consequently native animals that rely on native plants), is a poor filter of silt and pollutants, and grows so densely in wetlands that it crowds out fish and waterfowl. (Cindy Starr, "Purple Pest," *Cincinnati Post,* August 31, 1998)

In its native habitat, purple loosestrife only comprises one to four percent of the native vegetation, but in North America, densities of up to 80,000 stalks per acre have been recorded. (M. Strefeler, E. Darmo, R. Becker, E. Katovich, 1996, "Isosyme characterization of genetic diversity in Minnesota populations of purple loosestrife," *American Journal of Botany* 83:265–73)

The only creatures that eat this tall European flower [purple loosestrife] are European beetles; as a result the purple plague has overrun wetlands in 42 states from Maine to California, filled in open water and pushed several species of rare amphibians and butterflies to the brink. (Sharon Begley, "Aliens Invade America!," *Newsweek,* August 10, 1998)

A wetland with lots of purple loosestrife is soon a wetland with little wildlife. (City of Boulder Open Space Department, "Warning: Danger in the Field")

---

inant in ship ballast. Free from the insects and diseases of its homeland, purple loosestrife has a competitive advantage over native wetland vegetation and is now widely dispersed. It has spread rapidly in the Midwest in recent years, threatening not only native plants (including rare species) but also the wildlife that depend on them for food and shelter.

Once purple loosestrife is well established in an area, it can overrun wetlands and eliminate almost all other vegetation. Unfortunately, purple loosestrife is still promoted by some horticulturists for its beauty and by beekeepers as a nectar source.

Twenty-four states, including Indiana, Iowa, Minnesota, Missouri, Ohio, and Wisconsin, have laws to discourage the spread of purple loosestrife. In Wisconsin, for example, it is illegal to sell, distribute, plant, or cultivate purple loosestrife or any of its cultivars.

Habitat: Moist soil to shallow water (established plants will tolerate dry soil); sand to clay; sun to partial shade; can adjust to a wide range of growing conditions including flooding in up to 18″ of water; sometimes planted in gardens where it spreads to nearby wetlands, lakes, and rivers.

Height: 3–7′ or more.

Leaves: Simple; usually opposite on stems but sometimes alternate or bunched in whorls; lance-shaped; without petioles; edges are smooth; sometimes downy.

Flowers: Showy; individual flowers have five or six pink or rose-purple petals that are 0.5–0.75″ across and surround small, yellow centers; closely attached to the stem; bloom from the bottom of the flower spike up to the top from early July to September; plants can become quite large and several years old before flowering begins.

Seeds: Borne in capsules that burst at maturity in late July or August; main means of spread; single stems can produce 100,000 to 300,000 seeds annually; large, mature plants with fifty or more stems can release two million seeds each year; survival rate is 60 to 70 percent resulting in an extensive seed bank that remains viable for twenty years; can live twenty months submerged in water; most fall near parent plant; water, animals, boats, and humans assist in transporting them long distances; disturbances such as reduced water levels, damaged vegetation, or exposed soil increase germination; sterile hybrids sold by nurseries can pollinate wild loosestrife plants to produce viable seed.

Stems: 1–50 per plant; upright; stiff; usually four-sided (occasionally five- or six-sided); green to purple; often branching, making the plant bushy in appearance; somewhat woody; die back each year; clipped, trampled, or buried stems may produce new shoots and roots.

Root system: Consists of a large, woody taproot with fibrous rhizomes; rhizomes spread rapidly to form dense mats that aid in plant reproduction.

Similar species: Garden loosestrife (*Lythrum vulgaris*) is also an escaped nonnative purple loosestrife but not nearly as invasive. Both may be confused with the native wing-angled loosestrife (*L. alatum*) that is found in moist prairies and wet meadows. Native loosestrife has winged, square stems and solitary flowers in the leaf axils; lower leaves are paired, upper leaves alternate; generally a smaller plant than purple loosestrife. Purple loosestrife is also sometimes con-

Purple loosestrife is very showy with numerous, long flower spikes.

Flowers have 5–6 petals; bloom July to September.

Stems are 4-sided; leaves nearly linear, usually opposite; have smooth edges.

fused with fireweed (*Epilobium angustifolium*), blazing stars (*Liatris* spp.), and lesser purple-fringed orchid (*Platanthera psycodes*).

Control: Prevention is the most important element of purple loosestrife control. Plants in gardens should be removed. Wetlands should be checked annually for the presence of loosestrife, especially in July when flowering begins and plants are easy to spot. Ideally, since plants do not all flower at the same time, sites should be patrolled three times during the growing season, at one-month intervals after flowering begins. Isolated small colonies, especially in areas otherwise free of loosestrife, should be targeted for control first. Clean boots, shoes, and clothing carefully before leaving infested areas. Rinse all equipment, including boats and trailers, to prevent the spread of seeds.

If you can positively identify purple loosestrife, it is a good idea to start a control effort early in the growing season because plants may hold viable seed once flowering begins. Be careful not to leave loosestrife stems or cuttings behind since they can root or continue seed development. Control efforts should be completed before the earliest flowers, those lowest on the stems, disperse seeds. Follow-up control is critical for at least three years.

Manual or Mechanical Control: Small plants can be hand pulled. Older plants can be removed with a shovel, but avoid leaving behind any roots. Use caution when pulling or digging since disturbing the soil may encourage more seed germination. Removed plant parts should be landfilled, dried, and burned, or buried deeply in an area that will not be disturbed.

Mowing is not recommended because it can spread plant fragments and seeds, expose the seed bank to more sunlight, and lead to further seed germination. For limited control, cut budding stems at the base, and then bag them to make sure that seed development or spread does not occur. Continue this process because the plants will resprout. This method is of limited value because purple loosestrife will continue to spread by its root system.

Chemical Control: If applying herbicides near open water, a permit may be required. Check with the aquatic plant coordinator at your local natural resources department. Some states, like Minnesota, require posting "loosestrife control site" signs along public waterways or wetlands to notify anyone who may swim, fish, or boat in the area.

A selective herbicide control method involves applying a solution of 30% a.i. glyphosate to the raw area of freshly cut stems after the flowering portions have been removed. Treat all stems in a clump. This is generally the least destructive and most effective method for small infestations.

Currently, glyphosate formulated for use over water is the most commonly used chemical for killing purple loosestrife. Hand-spraying individual plants should be done in July or August. Use 1.5–3% a.i. glyphosate and an adjustable herbicide spray nozzle that will minimize overspray to nontarget plants. About 25 percent of the plant should be covered. Broadcast spraying of glyphosate is not recommended since it will kill too many nontarget plants. After spraying the herbicide, any purple loosestrife flower heads should be cut, bagged, and removed from the site.

Another type of foliar herbicide application, known as "the bloody glove technique," [K. Condict] involves using an absorbent cotton glove that has been sprayed with 5% a.i. glyphosate. The applicator, who wears chemical-resistant gloves on both hands with the cotton glove over one of them, works with another individual who cuts and bags the flowering portion of the purple loosestrife plant. The applicator then grabs the remaining plant with the treated glove about one-third the way down from the top and moves the herbicide-laden glove up the stem. In order to keep track of the treated areas, a dye should be added to the herbicide before work begins. Workers also wear chemical-resistant clothing when performing this type of application.

A 5% a.i. solution of triclopyr approved for aquatic conditions can be mixed with water and sprayed on loosestrife foliage. Unlike glyphosate, triclopyr will not harm monocots, such as grasses, sedges, cattails, rushes or reeds, when used according to label directions.

Biological Control: Biological control is currently considered the best option for large-scale control of heavy purple loosestrife infestations. Research involving the use of several of its natural insect enemies from Europe, especially *Galerucella* beetles, has been encouraging. Most of these insects feed almost exclusively on purple loosestrife and are little threat to native plants. Although they may eat native loosestrife, they prefer the exotic species. If handled properly, one hundred beetles can multiply to as many as ten thousand in just a few months. As the supply of purple loosestrife plants is depleted, beetles die or move on in search of more loosestrife.

Prevention, early detection, and control are still crucial in limiting the impact of purple loosestrife. If resources are adequate, and fast remedies are needed for small isolated populations of loosestrife, it is often best to use traditional methods of control to limit seed spread rather than waiting for beetle establishment. For more details on biological control, contact the aquatic plant coordinator at your local natural resources department. Most Midwestern states have programs that help people raise beetles for the biological control of purple loosestrife. See North American purple loosestrife control contacts, http:// misegrant.umich.edu/pp/nacontacts.htm.

Sweet Clovers
Yellow Sweet Clover: *Melilotus officinalis*
White Sweet Clover: *Melilotus alba*

Yellow and white sweet clovers are similar members of the Legume family. Native to Eurasia, these aggressive biennials degrade native grasslands by

Sweet clovers (white sweet clover, left, and yellow sweet clover, right) are biennials and can be controlled by well-timed spring burns.

shading out other species and outcompeting them for water and nutrients. They may also produce chemicals that inhibit the growth of other plants.

Sweet clovers were brought to North America in the late 1600s for forage and honey production. They later became popular for their ability to fix nitrogen in the soil, for production of rat poison, as an anticlotting agent in medicine, and for wildlife cover. Because they are economically and medically valuable, white and yellow sweet clovers will continue to be planted despite the problems they pose for land managers.

White sweet clover.

Yellow sweet clover.

Habitat: Direct sunlight or partial shade; prefer loamy or alkaline soils; will tolerate nutrient-poor soils; most frequently found in open, disturbed, upland sites such as prairies, savannas, and dunes; found in all fifty states but most frequently in the upper Midwest and Great Plains.

Height: 3–5′ in the second year; yellow sweet clover is usually shorter than white sweet clover.

Leaves: Compound; alternate; cloverlike; divided into three finely toothed leaflets with blunt or round tips; the middle leaflet grows on a distinct stalk above the other two; leaflet margins are toothed near the tips.

Flowers: Numerous; pealike; white on white sweet clover, yellow and blooms earlier on yellow sweet clover; 0.25″ long; densely arranged on irregular spikes located on the top 4″ of elongated stems; each small flower in a cluster is attached to the stem by a tiny stalk; bloom late May through September during the second year of growth; fragrant.

Seeds: The sole means of reproduction; up to 350,000 produced per plant; small; hardy; can remain viable in the soil for thirty years; burning breaks the hard, protective seed coat and stimulates germination.

Stems: Erect; many-branched; leafy; somewhat spreading at the base creating a bushlike appearance; often hollow.

Roots: Semiwoody; comprised of a deep taproot with extensive lateral roots.

Growth cycle: After germinating in late spring or summer, plants put their energy into developing healthy root systems the first year; roots survive the winter, and plants reappear in late April or early May of the second year, become large, produce flowers, set seed, and die.

Control:

Manual or Mechanical Control: Although burning stimulates sweet clover seed germination, burning two years in a row can greatly reduce populations. Burning should be conducted early the first year before green-up to stimulate germination. The area should then be checked in late summer for the presence of first-year plants. If found, another burn should be conducted the following year when plants are 6–8″ high. If flower buds have not yet developed, resprouting may occur. The second-year fire can be of low intensity—just enough to touch the stems. Damaged plants wither quickly, even if not completely destroyed by the fire. For small patches of sweet clover or in areas not completely burned, a propane torch can be used when vegetation is damp to avoid setting fire to the surrounding prairie. Heavily infested areas may need this burning sequence repeated after a few years to maintain control.

Small amounts of sweet clover can be controlled by hand-pulling first-year plants in late fall after root-crown buds have developed. Second-year plants can be pulled in May or June before flowering. Pulling is easier after rain or when the soil is moist. If pulling is done too early, many plants may be missed, and those with succulent stems may break off and resprout. Pulled plants will need to be removed from the natural area if pulling is done after seeds become viable. Inspect the site a couple of times during the summer for late-flowering plants that may have been missed.

If pulling plants is not an option, plants can be cut at ground level from the time flowering begins to full bloom using a brush lopper. Plants cut after full bloom should be removed from the site since they may hold viable seeds. In areas that have dense patches of sweet clover and few native plants, a power brush-cutter with a heavy-duty saw blade can be used. Check the site a week after cutting for plants that were missed.

Chemical Control: Sweet clover can normally be managed using manual or mechanical control and should not require the use of chemicals. If herbicide use is necessary, however, a good time to apply it is following an early spring burn, but before native plants emerge. Dicamba and clopyralid are effective on this species but should not be used in areas of sandy or loamy soil or where the water table is shallow. 2,4-D amine and metsulfuron-methyl are also effective.

Wild Parsnip
Pastinaca sativa

Wild parsnip often invades disturbed areas along roadsides.

Wild parsnip, a native to Eurasia, is an aggressive member of the Parsley family that frequently invades and modifies open, disturbed habitats. Once an infestation begins, wild parsnip spreads slowly across an area in waves. As the population builds, it begins to spread rapidly. A monocarpic perennial, each plant will die after successfully producing seed. Ancient Greeks and Romans ate its taproot, and cultivars are still grown today for food consumption.

CAUTION: Wild parsnip contains chemicals in the juices of its green leaves, stems, flowers, and fruits that can cause an intense, localized burn. Plant juices coming in contact with the skin while in the presence of sunlight can cause a rash, burn, or significant blistering. Resulting skin discoloration can last several months. To treat burns from wild parsnip, cover the affected area with a cool, wet cloth. If blisters are present, try to keep them from rupturing for as long as possible. The blister skin is "nature's bandage," keeping the skin below protected, moist, and clean while it heals. When blisters pop, try to leave the blister in place. To avoid infection, keep the area clean and apply an antibiotic cream. Adding aluminum acetate (e.g., Domeboro) powder to cool, cloth compresses can help dry, weeping blisters. Some doctors recommend a topical or systemic cortisone-steroid for extreme discomfort. For serious cases with extensive blistering, consult a physician.

Wild parsnip is a relatively tall herbaceous plant; blooms from early June to mid-July.

Habitat: Found in a wide range of growing conditions including dry, mesic, and wet-mesic prairies, oak openings, and calcareous fens; will not tolerate shade.
Height: 4′ or more.
Leaves: Plants remain in the rosette stage for one or more years; rosettes bear upright leaves averaging 6″ in height; mature plants have pinnately compound, alternate leaves; 5 to 15 oval to oblong hairless leaflets arranged in pairs along a common stalk; leaflets have saw-toothed edges; during the flowering stage when a tall stem emerges, leaves alternate up the stem, becoming smaller near the top.
Flowers: Yellow; small; five petals; hundreds per plant arranged in 2–6″-wide umbels at the tops of stems and branches; lateral flowers often rise higher than central flowers in the umbels; bloom normally occurs from the first of June to mid-July, although some plants may continue to flower through late summer.
Seeds: Fairly large; flat; round; slightly ribbed; yellowish; many produced on one plant; take at least three weeks to become viable after plant flowers; dispersal normally occurs from September through November; plants die after producing seed; seeds can remain viable in the soil for four years.

Leaves alternate on stem; leaflets arranged in pairs; have saw-tooth edges.

A member of the Parsley family, it has hundreds of tiny yellow flowers are arranged in umbels.

Stem: Single; erect; thick; hairy; grooved.

Taproot: Long; thick.

Similar species: Wild parsnip can be confused with the native prairie species, golden Alexander (*Zizia aurea*) and prairie parsley (*Polytaenia nuttallii*). Golden Alexander is typically shorter than wild parsnip, its umbels are less open, and it leaves have only three to seven leaflets. Prairie parsley leaves are compound, like those of wild parsnip, but its leaflets are oblong with few teeth. It has very small yellow flowers in somewhat rounded umbels, unlike the flat umbels of wild parsnip. Ensure identification before beginning control efforts.

Control: The best way to control wild parsnip is through early detection and eradication. Wear proper clothing in infested areas to avoid contact with plant juices (waterproof or heavy gloves with long cuffs, a long-sleeved shirt, long pants, socks, and shoes).

Manual or Mechanical Control: An effective control method for wild parsnip is to cut through the root 1 or 2″ below ground level with a sharp shovel before flowering has begun. Cutting below ground level prevents resprouting. With some soil types and wet conditions, plants can be pulled out by hand. If plants have flowered, and seed production has begun, seeds heads must be removed from the site. Burn them or bury them deeply in an area that will not be disturbed.

If the wild parsnip population is too large to hand-cut or pull, a power brush-cutter can be used to cut plants at the base of the stem just after peak flowering but before seeds set. When using a brush-cutter or string trimmer be sure to wear protective clothing, including a face shield, because these machines can spray bits of pulverized leaves and stems over any exposed skin, causing burns. Plants may resprout when cut above the ground and should be cut again a few weeks later to prevent reflowering. Mowing seems to encourage wild parsnip to send up a flowering stalk. It also sets back other plant species in the area that could offer competition. If used, mowing should be done before seed-set and repeatedly thereafter.

Cutting stems or hand-collecting seeds after seed-set, but before dispersal, reduces the likelihood that plants will be able to resprout and flower. Stems or seed heads cut at this time must be gathered and removed from the site to prevent seeds from maturing. If this control method is repeated over a number of years, the population will decrease as the seedbank is depleted.

Burning does not harm wild parsnip plants—they simply resprout. However, wild parsnip rosettes are easy to spot in the darkened soil following a spring burn and can then be dug out. Controlled burns may also stimulate growth in prairie species that compete with parsnip.

Chemical Control: Chemical control is effective but should be used sparingly in quality natural areas. Wild parsnip is one of the first plants to green-up following an early spring burn. Spot treatment with 1–3% a.i. glyphosate or metsulfuron-methyl plus a surfactant to basal rosettes at this time should not harm important native species that are still dormant. Triclopyr formulated for use with water and 2,4-D amine are also effective on this species.

Japanese Knotweed

Polygonum cuspidatum, syn. *Fallopia japonica* and *Reynoutria japonica*

Japanese knotweed resembles a shrub but has no woody tissue.

Japanese knotweed, also erroneously known as Mexican bamboo, is a semi-woody perennial in the Buckwheat family. It was introduced into North America as an ornamental plant from eastern Asia in the late 1800s and is, unfortunately, sometimes still planted for landscape screening and erosion control.

Japanese knotweed spreads rapidly by rhizomes forming large, dense thickets that eliminate native vegetation and wildlife habitat. The rhizomes are strong enough to damage pavement. Knotweed thickets are particularly problematic along waterways since they can limit access for fishing and wildlife, and cause flooding by decreasing water flow through rivers and channels. It is also one of the most troublesome weeds along railroad rights-of-way because it becomes a fire hazard during the dormant season. Once established, Japanese knotweed stands are extremely difficult to eradicate.

Habitat: Primarily along riverbanks and pond edges; also in wetlands and low areas, along hillsides, woodland edges, roadsides, and in yards; tolerates semi-shade and a wide variety of soils and moisture conditions; prefers moist, well-drained sites with nutrient-rich soils.

Height: To 10′.

Leaves: Simple; alternate; 3–4″ wide, 4–6″ long; egg-shaped to almost triangular; those on young shoots are usually heart-shaped; long petioles; broad, narrowing to a fine point at the tip; upper surface is dark green; lower surface is pale green.

Flowers: Creamy white or greenish; tiny (0.125″ wide); borne in plumelike clusters in upper leaf axils near the ends of stems; bloom August and September; male flower stems mostly erect; female flower stems droop.

Young plant; stems are hollow, bamboolike.

Plumelike flower clusters emerge from upper leaf axils; bloom in August and September.

Fruit: Small; triangular; shiny black; produced by female plant; rare since colonies seldom have both male and female plants.

Stems: Erect; bamboolike with arching branches; nodes are swollen and surrounded by a papery membrane; round; smooth; hollow; reddish brown; killed by frost but remain upright through the winter; new, reddish shoots emerge in spring among dead stems, which degrade slowly; dead stems and leaves create a thick groundlayer of organic matter that prevents establishment of native vegetation.

Root system: Comprised of fibrous roots and rhizomes that can reach 6′ deep and 65′ wide on parent plants; rhizomes are white when young, becoming brown, thick, and woody with age.

Reproduction: Spreads primarily by stout rhizomes and new shoots; does not appear to reproduce significantly by seed since fertile male plants are rare; new infestations often occur when contaminated soil from the use of earth-moving or road equipment is brought in;

colonies are also created when rhizomes wash downstream during flooding; young stems can produce new roots and shoots if buried in soil or floated in water.

Similar species: Giant knotweed (*Polygonum sachalinense,* syn. *Fallopia sachalinensis*) is very similar to Japanese knotweed but is much taller with larger leaves. The two species are known to hybridize. It is very invasive in the Northeast and is likely to become a problem in the Midwest.

Control: It is extremely important to begin control efforts as soon as a Japanese knotweed invasion is spotted. Areas downstream from established stands should be monitored for new infestations. Once established, Japanese knotweed is difficult to eradicate. Unless the stand is very small, eradication can take several growing seasons or may not be possible at all. Research is needed on the feasibility of introducing other plants for competition.

Manual or Mechanical Control: Although small plants can be dug or pulled out, this labor-intensive method of control is not generally recommended. New colonies can develop from cut stems or tiny rhizomes remaining in the disposed soil unless they are buried more than 3′ deep.

Cutting or mowing will reduce rhizome reserves, but this must be done at least three times during the growing season. Japanese knotweed, reportedly, does not survive weekly or bi-weekly mowing.

Covering colonies with black plastic may be helpful, especially if it is done in conjunction with cutting. Since Japanese knotweed is strong enough to penetrate asphalt, the plastic should be loose enough to allow for shoot growth beneath it, while continuing to keep the plant in dense shade. The plants can also be allowed to reach full height before being covered.

Chemical Control: Applications of 3% a.i. glyphosate with a surfactant, triclopyr formulated for use with water, or dicamba to the foliage may be effective on large populations, especially in early fall. Plants are more susceptible to herbicides if they are cut when 4–5′ tall, and the regrowth is then treated when about 3′ tall.

We predict that the invasion of non-native species will continue to be the greatest threat to the diversity of these [eastern United States] forests in the foreseeable future. (Peter M. Vitousek, Carla M. D'Antonio, Lloyd L. Loope, and Randy Westbrooks, *American Scientist* 84[5], September–October 1996)

Cut-stem treatment with a 30% a.i. solution of glyphosate or triclopyr formulated for use with water while plants are actively growing may also be effective.

Some managers report that foliar applications of 0.25% a.i. imazapyr have been effective for controlling Japanese knotweed. Treating cut stems with the same solution of imazapyr is also effective, but less so than the foliar treatment.

Tests involving large-bore needle injection of 5 ml of 41% a.i. glyphosate into one of the lower nodes of each stem in a given patch have been effective [J. Soll]. For more information, see the Web site, www.co.clark.wa.us/environ/knotweed.pdf.

Grasses and Grasslike Species

Two species discussed below, reed canary grass (*Phalaris arundinacea*) and common reed grass (*Phragmites australis*), are thought to be native to much of North America, including the upper Midwest, as well as other parts of the world. Some populations, however, are very aggressive and are believed to be introduced varieties.

Reed Canary Grass
Phalaris arundinacea

Few plants can grow in areas dominated by reed canary grass.

Reed canary grass is a sod-forming, cool-season perennial grass that is native to the temperate regions of Eurasia and North America. Several cultivars or subspecies, selected for their vigor and possibly Eurasian in origin, have been planted throughout the United States since the 1800s for forage and erosion control. While there is currently no reliable way to tell the genetic origin of an individual stand, most land managers believe that the majority of Midwestern reed canary grass colonies are Eurasian. It has become naturalized in much of the northern half of the United States and southern Canada, and is still being planted on steep slopes, banks of ponds, and in created wetlands. Agricultural cultivars are widely available for wet sites and for upland hayfields.

Over time, reed canary grass can form large colonies that harbor few other plant species and are of little use to wildlife. It typically spreads throughout a wetland or floodplain and is increasingly being found invading forested sites, limiting tree regeneration. Because selective control is difficult, this highly adaptive species is of particular concern.

Reed canary grass prefers moist soil but can also grow in upland sites.

Habitat: Ditches, streambanks, wet meadows; does best on disturbed, fertile, moist, organic soils in full sun; can invade most types of wetlands, including lowland forests; threatens wetland sites undergoing restoration; can also grow on dry soils in upland habitats and in partial shade.

Height: 2–5′.

Leaves: Blades are flat; rough texture on both surfaces; gradually tapering from base to tip; top leaves are horizontal; sheath is smooth and rounded; prominent ligule, long and thin; no auricles.

Seed heads: Green to purple when in bloom (normally May to mid-June); change to golden tan as seeds form; flower branches are spreading during bloom stage but draw close to the stem at maturity; 3–16″ long with lateral branches 0.5–1.5″ in length.

Seeds: Shiny brown; ripen in late June and shatter; dispersed from one wetland to another by waterways, animals, humans, and machines; seeds can germinate immediately at maturation.

Bloom occurs May to mid-June.

Closeup of inflorescence in early summer.

Stems: Erect; hairless; hollow; smooth and rounded.

Root system: Short, stout rhizomes that root at the nodes forming a thick fibrous root mass.

Reproduction: By seed and creeping rhizomes.

Life history: One of the first grasses to sprout in the spring; uses energy to produce leaves and flower stalks for five to seven weeks after germination, then spreads laterally; stems collapse in mid- to late-summer and form a dense, impenetrable mat with the leaves. Plants tend to form clumps 3′ or more in diameter.

Similar species: Reed canary grass resembles non-native orchard grass (*Dactylis glomerata*) but is generally taller, has wider leaf blades, and narrower, more

Reed canary grass is one of the first grasses to emerge in spring.

Panicles become compact, beige by midsummer.

pointed inflorescences. Bluejoint grass (*Calamagrostis canadensis*), a native species, may be mistaken for reed canary grass in areas where orchard grass is rare, especially in the spring. The highly transparent ligule of reed canary grass is helpful in distinguishing it from other species. There is another European canary grass (*P. canariensis*) that is naturalized in North America, but it is much less abundant.

Control: Stands of reed canary grass are difficult to eradicate because their huge seedbank will recolonize the site. No single control method works everywhere. In native plant communities, mechanical control methods are recommended. In buffer areas and on severely disturbed sites, chemical and mechanical controls may be used. Any method to reduce or eliminate reed canary grass should be followed by planting native species appropriate for the site, if they are no longer present, to offer competition.

Manual or Mechanical Control: In the early stages of reed canary grass invasion, small patches can be dug up or covered with black plastic for at least one growing season. Bare spots can then be reseeded with native species. Covering with plastic is not always effective since rhizomes can spread beyond the plastic edges.

Close mowing three times a year can help control reed canary grass. It should be done in spring to retard growth, at flowering to prevent seed formation, and in late fall to repress growth for the following year. Mowing also exposes the ground to light and promotes the growth of native species for competition.

In badly degraded areas, Bobcats or other types of machinery have occasionally been used to remove reed canary grass and the underlying soil. This not only eliminates the reed canary grass but also removes the seedbed and the silty sediments that reed canary grass is often associated with. This has been especially effective in wetlands that have been artificially drained by ditches. The reed canary grass sod and the first 8–12″ of soil is pushed into the ditches, both restoring the hydrology of the site and exposing the peaty soil for recolonization by native species.

Grazing may help set back reed canary grass, but cattle will not always eat it. Grazing is most effective in midspring though early summer.

For regenerating trees in reed canary grass infestations, some restorationists have had success planting cuttings of willows, which compete well with the reed canary grass and shade it out after a few years. The desired trees are then planted among the willows.

Integrated Control: When using an herbicide for the following control methods, buy one approved by the Environmental Protection Agency for use in areas of standing water or for aquatic applications. Permits from your natural resources department will also likely be necessary for herbicide applications in these areas.

Repeatedly disrupting roots weakens plants and depletes the seedbank. In small areas with few native plants, reed canary grass can be tilled every two to three weeks for one full growing season followed by dormant seeding with native species near the first frost date. When combined with spot applications of 3% a.i. glyphosate in areas too wet for early or late cultivation, results after two years have been good. (Use a wick for herbicide application rather than spraying to limit contact with nontarget species.) Frequent cultivation is important since one or two cultivations will simply cut roots up and increase the number of individual plants.

Controlled burns in late spring or late fall for five to six years may help reduce reed canary grass populations. Fires, however, are often difficult to conduct due to water levels and the greenness of grass at the time of burning. An application of 3% a.i. glyphosate will brown-off reed canary grass enough to conduct a controlled burn. A late-spring burn, followed by mowing or a wick application of glyphosate to emerging shoots, will eliminate seed production for that year. Burning does not work in dense stands of reed canary grass that lack competition from native, fire-adapted species.

Chemical Control: Small (2′ in diameter), scattered clones of reed canary grass can be controlled by tying stems together just before flowering, cutting off and bagging the inflorescence, and applying 20% a.i. glyphosate to the cut stems.

Since reed canary grass begins growing early in the spring, a 5% a.i. solution of glyphosate can be applied to young shoots before most native plants emerge. Dead leaves from the previous year should be burned or removed first so shoots have better exposure to the herbicide.

Mowing in mid-September, followed by an application of 5% a.i. solution of glyphosate in October, when most native plants are dormant, can also help control this species. (Fall applications of glyphosate may be more effective.)

In dry areas, grass-selective herbicides, such as sethoxydim or fluazifop-p-butyl, can be used to treat large patches of reed canary grass early in the season before native grasses emerge, or in the fall after most native grasses have gone to seed. If necessary, replant the area with native sedges and wildflowers to offer competition. (Sedges will not be killed by these grass-specific herbicides.)

Of all the herbicides, imazapic is the best to use for long-term control of reed canary grass. Be aware, however, that it may harm sedges.

Common Reed Grass

Phragmites australis, syn. *P. communis*

Common reed grass, a natural part of many wetland plant communities, is a tall, warm-season perennial grass with featherlike plumes at the top of its stems. It has been used by various cultures to provide numerous products including fiber for mats, roof thatching, fishing rods, mouthpieces for musical instruments, paper pulp, insulation, and fuel.

The stout rhizomes of common reed grass help stabilize shorelines, and winter stalks provide cover for wildlife, such as deer, pheasant and rabbit, but invasive stands can become so dense that diverse wetland plant communities are eliminated and navigation becomes difficult. Large colonies can also block irrigation systems and cause flooding.

Although common reed grass is native to much of the world, including the upper Midwest, highly aggressive strains that form dense, impenetrable stands with few native species are believed to be non-native in origin. Genetic research has shown that non-native stands were introduced on the Atlantic Coast in the early twentieth century and have been spreading across the continent. There are a number of field characteristics that can be used to distin-

Common reed grass forms large colonies because of its spreading rhizomes.

Common reed grass has a plumelike inflorescence and large, horizontal leaves.

guish between native and non-native strains. These are detailed on the following Web site: www.invasiveplants.net/. The characteristics listed below are typical of the non-native strains, although diagnosis should involve a suite of characteristics, not just a few.

> *Habitat:* Roadside ditches, open wetlands, riverbanks, lake shores, disturbed or undisturbed plant communities; prefers alkaline and brackish waters but will tolerate highly acidic conditions; can grow in water up to 6′ deep and in somewhat dry sites.
> *Height:* 3–20′.
> *Leaves:* Linear; green or grayish green; 0.5–1.5″ wide at the base; 10–20″ long; smooth; flat; rough on the margins; leaf sheaths stay attached to the stem through the winter.
> *Seed heads:* Large, dense, featherlike, grayish purple plumes; 5–16″ long; produced in late July through September; become beige to dark brown at maturity.
> *Stems:* Canelike; up to 1″ in diameter; non-native strains are tan, dull, rough, and ribbed, rigid and tough; hollow; unbranched; buds form on the rhizomes during the summer and become fast-growing stems the following spring.
> *Root system:* Rhizomes that can reach up to 6′ deep with roots emerging at the nodes.
> *Reproduction:* Large colonies are formed by spreading rhizomes; stolons sometimes contribute to expansion; seeds are

Common reed grass in fall. Plants can grow more than 8′ tall.

often infertile; non-native strains have rapid clonal expansion, while native strains generally do not appear aggressive.

Control:

Manual or Mechanical Control: Simply mowing common reed grass has not proven effective for control. Although mowing may reduce the overall vigor of this species, it adds a substantial amount of organic matter to the system, involves a lot of trampling, and reportedly does not reduce the size of the infestation.

Digging the massive common reed grass rhizomes is very difficult. Disposal of soil contaminated with viable root fragments or seeds also becomes a problem, making this method of removal unappealing. However, using a bulldozer to scrape off the plants and rhizomes and then dump them into drainage ditches may be effective in wetland restorations.

Long periods of flooding during the growing season might be considered a control option if water levels can be manipulated, and all or most of the plant is inundated. Older stems should be burned, mowed, or crushed to a point below the anticipated water level before flooding. Consider the impact on native species before using this technique.

Chemical Control: Permits from your natural resources department will likely be necessary for use of herbicides in areas of standing water. Herbicides used in these areas should be approved by the Environmental Protection Agency for aquatic applications.

Various techniques have been used to control common reed grass with herbicides. Before choosing an application method, however, consider the density of the infestation. Backpack sprayers are typically used to treat dense monocultures of this species, since there are few native plants likely to be harmed by overspray. Using a wick applicator, on the other hand, is more appropriate in areas where native vegetation persists within a common reed grass clone. Adding an appropriate dye to the herbicide before application will help the applicator keep track of the treated areas.

For large stands of common reed grass, glyphosate with 1.5% a.i. can be applied to the upper foliage in early fall (after seed heads are fully emerged) with a backpack sprayer and a 5′ wand extension. Because seeds seldom germinate, seed heads do not have to be removed. The area can then be burned the following spring to remove the large mat of dead plants. This will make a follow-up treatment to resprouts in early June easier and allow the

native seed bank to germinate. A 97 percent mortality rate has been reported after using this method for one year [O'Leary]. This process may need to be repeated after several years to maintain control.

Glyphosate with 1.5% a.i. can also be sprayed on the foliage of common reed grass resprouts in September after mowing the infestation in early August. Resprouts should be from knee- to waist-high for convenient herbicide application.

For small infestations, common reed grass can be clipped by hand in early August followed by an application of glyphosate (1.5% a.i.) to the resprouts of individual stems. To apply the herbicide, chemical-resistant gloves should be worn with a cloth glove over the top of one. The cloth glove is then carefully sprayed with herbicide and used as a wick to treat the stems.

Another method used for small infestations involves hand-cutting stems near the ground in July or August followed by dripping glyphosate into the cut stem. Using 25% a.i. glyphosate for this method has resulted in a 50 to 75 percent mortality rate [O'Leary], while using 50% a.i. glyphosate has resulted in little resprouting two years later [Morris]. Treating each stem immediately after cutting will help insure that all stems receive herbicide applications.

With small but dense infestations, gather, bundle, and tie an armload of stems together. Cut the stems above the string and spray the cut stem tops with one of the herbicides listed above.

Imazapyr is also reported to effectively control this species.

Cattails

Narrow-leaved cattail: *Typha angustifolia*
Hybrid cattail: *T.* x *glauca*

Cattails are familiar aquatic perennials that provide food and shelter for many marsh-dwelling animals. Their shoots and rhizomes are a basic part of the muskrat diet, and submersed stalks provide shelter and spawning habitat for fish. Cattails are also important for stabilizing shorelines and removing water pollutants. Various cultures have used cattail rhizomes for food, its leaves to make mats, baskets, paper, and other items, and its seed head fluff for insulation and stuffing in life preservers.

Three cattail species are found in the upper Midwest. Common cattail (*Typha latifolia*) is native throughout North America and Eurasia. Narrow-leaved cattail

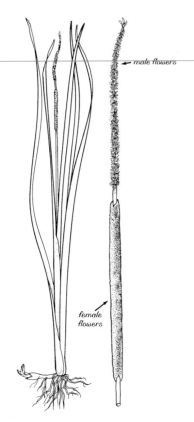

Cattails flower heads have both female and male flowers. The female flower section is brown-colored at maturity and located below the male flower section, which reaches to the tip of the flower head.

(*T. angustifolia*) is probably native only to Eurasia but is now established throughout much of the United States. It is abundant in the Midwest, where it hybridizes with common cattail to produce the mostly sterile "hybrid cattail" (*Typha* x *glauca*). Southern cattail (*T. domingensis*) is native in the southern Midwest, where it forms fertile hybrids with narrow-leaved cattail. Although the narrow-leaved and hybrid cattails are considered ecologically invasive, common cattail sometimes needs control as well to promote diversity in disturbed areas.

The amount of acreage in the Midwest dominated by cattails has increased dramatically since the early twentieth century due to wetland habitat modification by humans and the spread of narrow-leaved cattail westward from the Atlantic Coast. Cattails can outcompete other wetland vegetation to form dense monocultures in which dense rhizomes, leaves, and stalks, reduce overall habitat value. Many wetland areas, which were once havens for waterfowl and wading birds with a mix of cattails, open water, and diverse plant life, are now solid stands of cattails in which few species can live.

Note: Common cattail, a plant native to the upper Midwest and very similar to the two species, is discussed here rather than in chapter 6, "Native Plants That Sometimes Need Control," for the reader's convenience.

Habitat: Wetlands, lakeshores, river backwaters, roadside ditches, disturbed wet areas, consistently damp patches of rural and suburban yards; areas with wet soil or emergent in 3–4′ of water; in nutrient rich or slightly saline soils. Narrow-leaved cattail and hybrid cattail are more abundant than common cattail in places with more siltation and higher levels of nutrients or salt.
Height: 4–12′.
Leaves: Long; graceful; swordlike; spongy; veins are parallel; can be 3′ tall; originate at the base of stems and spread outward as they rise into the air; contain hollow chambers in cross-section.

> *Narrow-leaved cattail:* 0.25–0. 5″ wide; dark green; rounded on the back; top of the leaf sheath has thin, ear-shaped lobes at the junction

Narrow-leaved cattail has a 0.4–0.5″ gap between the male and female flowers (note where male flowers were attached) and dark green leaves with rounded backs.

Hybrid cattail has a 0.2–2″ gap between the male and female flowers, a longer and thicker female flower section, and longer leaves.

with the blade that usually disintegrates in summer.

Hybrid cattail: 0.4–0.6″ wide; often has taller leaves than the parent plants; top of the leaf sheath has thin, ear-shaped lobes at the junction with the blade that often disintegrate in summer.

Common cattail: 0.4–0.8″ wide; top of the leaf sheath has thin, ear-shaped lobes at the junction with the blade that persist in summer.

Flower head: A velvety brown, hot dog–like spike located at the ends of stem; male portion above the female portion; matures by midsummer.

Narrow-leaved cattail: Has a 0.4–0.5″ gap of bare stem between the male and female flowers; the female spike is 4–8″ long and 0.5–0.9″ thick.

Hybrid cattail: Sometimes has no gap, but usually has a gap of from 0.2–2″ of bare stem between the male and female parts of the flower head; female spikes are often longer (up to 15″) than the other two species and are 0.6–1.2″ thick.

Common cattail: Has no gap between male and female portions of the flower head.

Seeds: One seed in each of the single, tiny fruits in each flower; one plant can produce 250,000; wind-dispersed; some are released in fall, others remain on the spike until spring; remain viable in the seedbank for up to one hundred years.

Reproduction: New colonies develop from seeds on bare mud or in very shallow water; seedlings emerge the second growing season and spread rapidly by means of thick, starchy rhizomes to form large clones. Muskrats and storms often break off pieces of rhizomes, which are then dispersed by water to form new colonies. Hybrid cattail produces few seeds but spreads primarily by rhizomes.

Similar species: When not in flower, the following species might possibly be confused with cattails: blue flag iris (*Iris versicolor*), which has blue-green, flat leaves, and sweetflag (*Acorus calamus*), which has leaves with an off-center midrib and spicy smell when crushed.

Control: Management of cattail populations should be site specific and will depend on the hydrologic state of the site, the size of the area to be managed, and whether water levels can be manipulated. Seek the advice of experts in your natural resources department, and obtain the most current control regulations before beginning cattail management. One or more permits may be necessary.

Although eliminating only the narrow-leaved and hybrid cattails from high-quality natural areas should be considered, in most situations cattail populations are too large, the three species too intermingled, and differentiating between the species too difficult to make this option feasible. In general, the goal of cattail management should be to control their spread and density, not total eradication. Monotypic stands should be discouraged. A mix of 50 percent vegetation and 50 percent open water will generally allow for species diversity and wildlife habitat.

Manual or Mechanical Control: High water levels and flooding can be very helpful in controlling cattails and preventing seed germination. With the techniques that follow, however, be aware that cattails are not always set back by high water levels. They will sometimes float to the surface as a mass of entwined rhizomes and dead leaves, and continue growing as a floating mat until water levels return to normal.

Both living and dead cattail leaves provide oxygen for their rhizomes. To suffocate cattails and prevent air from reaching the root system, dead leaves should be cut or crushed in late summer or early fall far enough below the water surface so that what remains will be covered by water through the winter. Leaves can also be removed in early spring and water levels raised to keep the shoots covered. If cattail shoots are allowed to penetrate the water surface, air will once again be restored to the rhizomes. If shoots are to remain submerged, the eventual depth of the water necessary to kill cattails will vary depending on the average temperature (cattails grow more vigorously in hot weather) and the amount of starch that the plants stored in their rhizomes the previous year. A water level maintained at 3–4′ above the tops of the shoots will generally retard their growth. It may take two years of high water levels before a cattail population declines.

Maintaining a water level at 4–5′ above shoots favors the survival of muskrats. They harvest numerous cattail leaves to make their houses, and eat the shoots and rhizomes as the core of their diet. Population levels of ten muskrats per acre, combined with high springtime water levels, can nearly eliminate the emergence of cattails within two years. Overtrapping of these important animals should be discouraged whenever possible.

Cutting or burning cattails during a drawdown period, then restoring water to normal levels has also been effective in achieving control.

Grazing by cows, geese, and other animals can be helpful in managing cattail populations as well, especially when flower spikes are emerging.

Although frozen ground or saturated soil can impede a fire's progress through cattail duff, cattail marshes can often be burned in winter or when new shoots are coming up in spring. These are generally the only times when fuel is dry enough to carry a fire. Burning in spring, when cattail growth is just beginning, stresses the plants since carbohydrates from the roots are going to the leaves at that time. It is most effective if high water levels follow. If water levels cannot be manipulated and rain does not flood the area, many cattails will likely return. Although using fire for cattail control will release ash and some nutrients to the natural area, it will also liberate nutrients to the atmosphere and consume some of the mat, which will discourage cattails from floating.

During times of drought, cattail stands overlying well-developed peat soils can also be eliminated by burning. Because such fires burn the peat as well land managers must be able to smother the fire by reflooding the marsh. Peat fires can also cause undesirable changes in the marsh environment, including destruction of the seedbank, loss of peat, and air pollution.

If mechanical means are being considered for cattail control, such as shearing with the front-end loader on a tractor, disking, aerial spraying, mowing, bulldozing, or crushing with a weighted drum pulled by an all-terrain vehicle, contact your natural resources department for more information because permits may be required.

Chemical Control: Check with your natural resources department regarding posting and licensing requirements for herbicide use. Selective wick, boom, or hand-spray applications of glyphosate to cattail foliage in mid- to late summer, followed by cutting and removing dead stems a week later, may be effective. When working in areas of standing water, use glyphosate approved by the Environmental Protection Agency for aquatic applications. Annual chemical treatments will likely be necessary for a few years due to the massive root systems. Aerial herbicide applications may be an option on large cattail sites, but they may be costly and care must be taken to avoid impact on nontarget species. Imazapyr, 2,4-D, and other herbicides are also used to control cattails.

Aquatic Plants

Eurasian Water-Milfoil
Myriophyllum spicatum

> **Eurasian water-milfoil**
>
> In 1993, 36 percent of the boats and trailers exiting lakes in Hennepin County, Minnesota, were found to be contaminated with fragments of Eurasian water-milfoil. In 1996, $146,000 was spent for the control of Eurasian water-milfoil in forty-eight Minnesota lakes. (Minnesota Department of Natural Resources)

Eurasian water-milfoil in September.

Eurasian water-milfoil is an herbaceous submerged aquatic plant, native to Eurasia and northern Africa. It is the only non-native water-milfoil abundant in most of the upper Midwest. (*Myriophyllum aquaticum,* another non-native, is found in Ohio and may have the potential to spread in other parts of the Midwest.) There are eleven native water-milfoil species in the United States. Its ability to stay alive over winter, begin growing rapidly in spring, and block out sunlight needed by important native plants often results in its total domination of an area. Infestations of Eurasian water-milfoil threaten aquatic plant communities by keeping out larger fish, which disrupts the predator-prey relationship. Eurasian water-milfoil also impairs the ability of some fish to spawn and reduces the number of nutrient-rich native plants available for waterfowl. Its dense mats also affect the aesthetics, water quality, and recreational opportunities of lakes and slow-moving rivers and streams.

The feathery leaves of Eurasian water-milfoil whorl around the stem.

Habitat: Usually found in water 3–12′ deep; grows best in fertile, fine-textured, inorganic sediments and alkaline systems with a high concentration of dissolved inorganic carbon; prefers highly disturbed lake beds, heavily used lakes, and lakes receiving nitrogen- and phosphorous-laden runoff; often becomes the dominant species in lakes laden with nutrients; also found in ponds, slow-moving streams, reservoirs, and estuaries; high water temperatures allow it to flower and fragment repeatedly; is not as successful in undisturbed areas where native plants are well established.

Leaves: Submerged; feathery; limp when out of water; 4–5 leaves whorl around the stem at each node; typically uniform in diameter; consist of 9–21 threadlike pairs of leaflets that collectively resemble bones on a fish spine.

Flowers: Tiny; inconspicuous; located in the axils of floral bracts; either four-petaled or without petals; flower spike rises 2–4″ above the water surface.

Fruit: Four-jointed, nutlike bodies.

Stems: Slender; thicken below the flowers; double in width farther down; become leafless near the base; usu-ally 3–10′ long, but can reach 33′; often branch repeatedly at the water surface creating a canopy of floating stems and leaves.

Reproduction: Primarily by fragmentation due to boating and wave action; plants also fragment once or twice a growing season on their own; new plants emerge at leaf nodes and break away after flowering and fruit production; when plants float to new areas, roots that developed earlier along the stems are ready to take hold in the sediment and start new colonies; boats, motors, trailers, bilges, live wells, and bait buckets can unintentionally spread fragments over great distances and from one water body to another; fragments stay alive for weeks if kept moist.

Eurasian water-milfoil also reproduces by runners that creep along the lake bed; runners, lower stems, and roots survive over winter and store carbohydrates that help it grow fast early in the spring. Seeds typically germinate poorly under natural conditions.

Similar Species: Without flowers or fruits, Eurasian water-milfoil may be difficult to distinguish from the native, northern water-milfoil (*M. sibiricum*). Northern water-milfoil leaves typically have seven to eleven pairs of leaflets and remain rigid out of water. It also develops winter buds (Eurasian water-milfoil does not), is less aggressive, and seldom forms a branched canopy on the water's surface. Coontail (*Ceratophyllum demersum*), often mistaken for the water-milfoils, has forked leaves rather than the featherlike leaf divisions of water-milfoil.

Control: Permits from your natural resources department will likely be required for chemical treatments, bottom screening, buoy/barrier placement, and some mechanical harvesting devices.

To prevent the spread of Eurasian water-milfoil:

- Boaters should remove all aquatic vegetation from their boats and equipment before leaving any lake or river landing. Dispose of them in garbage cans at the boat access area or at home.

- Drain all bilges, live wells, and other water containers before leaving the water access area.

- Do not transfer water from one water body to another.

- Wash your boat and trailer thoroughly with regular tap water when you get home. Flush water through your motor's cooling system, live wells, and other areas that hold water. Preferably, dry your boat and equipment for three days in a sunny location before transferring it to a new body of water.

- Boaters, jet skiers, and mechanical plant-removal operators should avoid beds of emergent and submergent native plants.

- Buoys can be used to mark Eurasian water-milfoil colonies and warn boaters to stay away.

- Watershed management programs should try to prevent nutrients from reaching lakes.

- New infestations should be controlled immediately and monitored to minimize their spread.

Manual or Mechanical Control: Hand pulling is the preferred control method for Eurasian water-milfoil colonies less than 0.75 acres or areas with fewer than one hundred plants. Raking is also an option. All plant fragments must be removed from the water and shoreline area. (Water-milfoil can be used as a garden mulch.)

Sites far from boat traffic can be covered with bottom screens that are anchored firmly against the lakebed. These screens create a physical barrier to growing plants. Plants are unable to push their way through the barrier and eventually die. Screens can be expensive and should be used only for severe infestations, since they will kill native vegetation as well. They must be removed periodically to clean off sediment where Eurasian water-milfoil fragments can root.

Mechanical harvesters should be used only after colonies have become widespread. Besides creating fragments that lead to more Eurasian water-milfoil plants, they remove beneficial aquatic vegetation. Harvesters should be used offshore, in water at least 2–3′ deep, where they have room to turn around. Hand-cutters can be used inshore along with hand pulling and bottom screening.

Manipulating water levels may be helpful where possible. Low water levels can dry up Eurasian water-milfoil plants if maintained for several months, but will also destroy native plant communities. High water levels can drown Eurasian water-milfoil by limiting access to light.

Dyes are also sometimes added to water to prevent light from reaching Eurasian water-milfoil plants. The impact on native plants should be considered before using this method since they will also receive less light.

Replacing Eurasian water-milfoil with native plants will give new invasions competition, stabilize sediments against wave action, and provide habitat for wildlife.

Chemical Control: Herbicides for aquatic plant control should be used with utmost care. The technology for herbicide control of aquatic plants has been evolving rapidly in recent years. Aquatic experts should be consulted for recent findings before beginning an aquatic management program. Any control plan should be formulated on a site-specific, goal-specific basis. Herbicide use on some sites may not be locally acceptable if the herbicide is perceived as a greater risk than aquatic weeds.

2,4-D and fluridone are the most commonly used herbicides against Eurasian water-milfoil. Both chemicals have had mixed results and can affect native plant populations as well as temporarily restricting water usage for humans. Fluridone is a selective, systemic herbicide that can be applied in low doses to Eurasian water-milfoil in late fall. Applied at this time, it will affect perennials that overwinter. 2,4-D is applied in early spring or late fall. Timing and correct dosage are important for both of these herbicides to be selective in their impact. Adding surfactants and other chemicals may increase herbicide toxicity. The aquatic version of triclopyr has also proven itself effective for controlling Eurasian water-milfoil. Rapid water movement or any condition resulting in rapid dilution of these products in treated water will reduce their effectiveness.

Biological Control: Researchers have found a native North American weevil (*Eurhychipsis lecontei*) that feeds on Eurasian water-milfoil. Overall, results have been mixed with some sites showing significant fluxes in milfoil populations. Their use at this time is generally limited to experimental studies. For more information, contact your aquatic plant coordinator at your natural resources department or call Enviroscience, Cuyahoga Falls, OH (800) 940-4025 or (330) 940-4300.

4

Invasive Plants of Lesser Concern

THE PLANTS DISCUSSED in this chapter have invaded natural areas in various parts of the upper Midwest and collectively affect a large area. Large populations of these plants sometimes exist at the local level, especially in disturbed areas. The plants here do not pose a major threat to well-established native plant communities, although they might pose problems for areas being restored. Others are widespread and abundant but generally do not displace native plants. Control recommendations for some species may be unavailable or inconclusive. See chapter 2, "Invasive Plant Control Techniques," and Appendix A, "Resources for Information about Invasive Plants," for additional helpful information.

TREES AND SHRUBS

Norway Maple
Acer platanoides

A large, deciduous tree from Eurasia that is widely planted as a street tree and is now often considered overplanted. Its dense canopy shades out native wildflowers. It can outcompete, and is often mistaken for, the native sugar maple (*A. saccharum*). Sugar maple does not have white sap in the leafstalks or buds, and its fruits (samaras) are somewhat horseshoe-shaped and smaller (1–1.75″ long) than Norway maple. Many cultivars of Norway maple are commercially available.

Norway maple

> *Habitat:* Disturbed forest communities in or near urban areas; very shade tolerant.
> *Height and Form:* 40–60′ tall; spread is two-thirds of its height or equal to it; rounded, symmetrical crown; dense canopy; 1–3′ diameter trunk.
> *Leaves:* Simple; opposite; 3–7″ wide; toothed; 5–7 pointed lobes; shiny dark green above; shiny below with hairs occasionally in axils of veins; yellow in fall (many cultivars have varying colors in summer and fall); leafstalk is 3–4″ long and produces milky sap when removed from stem; leaves remain on trees late in season.
> *Buds:* Large; rounded; green or reddish; one large bud in center with two smaller lateral buds; large end bud will emit milky sap if cut and pressure is applied.
> *Flowers:* Yellow; in clusters; bloom in May; develop with the leaves.
> *Fruit:* Double samara with horizontally spreading wings; 1.5–2″ long; one seed in each half.
> *Twigs:* Stout; smooth; olive to gray-brown.
> *Bark:* Grayish black with shallow furrows.
> *Reproduction:* By seed.
> *Control:* Seedlings can be pulled or dug when the soil is moist. Girdling will control as well as cut stump or basal bark herbicide treatments.

Tree-of-heaven can grow 80′ tall.

Tree-of-Heaven
Ailanthus altissima

A fast-growing deciduous tree from eastern and central China, often planted in urban areas due to its tolerance of pollution and poor soils. It is becoming a serious problem in forests in some eastern states. Saplings can grow 3–4′ a year and form thickets that outcompete native vegetation for sunlight. It is also allelopathic. Can be an agricultural pest with hundreds of seedlings springing up in newly planted fields; also spreads rapidly when forest canopies are opened up. Its wood is weak and rots quickly, and its leaves and flowers cause contact dermatitis in some people.

Habitat: Disturbed soils, fence rows, fields, roadsides, woodland edges, forest openings, rocky areas; thrives in poor soils and tolerates pollution; not found in wetlands or shade.

Height and form: Small to large, but can grow up to 80′ tall; 1–2′ diameter trunk.

Leaves: Pinnately compound; alternate; large; 1–3′ long; 11–30 lance-shaped leaflets per leaf, leaflets are lance-shaped and 3–5″ long, less than 2″ wide; entire—except for 1–5 small gland-tipped teeth near the base; dark green above, pale green beneath, turn yellow in fall; have unpleasant odor resembling rancid peanut butter when crushed.

Flowers: Small; yellow-green; 5–6 petals; borne in dense clusters near ends of upper branches in late spring; male and female flowers on different plants; pollen of male flowers has offensive odor.

Fruit: Pinkish to tan; papery; flat; winged with single seed in middle; develop in clusters on female trees in fall; may remain on tree through winter.

Stems: Smooth.

Bark: Gray to brownish gray; nearly black with age; thin; twigs light to dark brown; large leaf scars on branches in winter are arranged in a V-shape.

Root system: Aggressive; spreading rhizomes may damage sewers and foundations.

Reproduction: By seed and shoot production; up to 350,000 seeds produced annually by a single plant; seeds germinate readily; spread by wind, birds, and water.

Control: Seedlings can be pulled when the soil is moist, but all root fragments must be removed. If infestations are small, it may be possible to kill this species by repeatedly and frequently cutting all stems in a clone close to the ground. This will likely be necessary for several years until root stores are depleted. Be prepared to control suckering. Infrequent or a single cutting of stems without the application of herbicide, such as glyphosate or triclopyr, to all cut stumps in a clone will lead to vigorous resprouting and root suckering. Foliar treatments of glyphosate or triclopyr have been effective as long as trees are small enough to adequately spray the foliage without harming

Pinnately compound leaves can be 1–3′ long with 11–30 leaflets; note the distinctive notches near the base of each leaflet.

Pinkish fruits develop in fall.

desirable plants. Basal bark treatment with triclopyr has also been successful. Applications have reportedly been most productive when made in late winter or in summer; fall to midwinter applications have not had good results. Follow-up monitoring and treatment for suckers and root sprouts will likely be needed for continued control.

European or Black Alder
Alnus glutinosa

European alder is a quick-growing, midsize tree, native to Europe, western Asia and northern Africa. Planted as an ornamental tree, there are several cultivars or varieties.

European or black alder has conelike fruits (catkins) and blunt-tipped leaves.

> *Habitat:* Along rivers, ponds, or wetlands; forests; prefers wet soils and full sun but is adaptable to poor or dry soils, and soils of various pH.
>
> *Height and Form:* 30–50′; pyramidal form when young, becoming ovoid or oblong and irregular with age; 1–2′ diameter main trunk; may have several smaller diameter secondary trunks.
>
> *Leaves:* Simple; alternate; 2–4″ wide by 2–5″ long; toothed (usually double); rounded; blunt-tipped; dark green above, lighter below; young leaves gummy to touch.
>
> *Flowers:* Monoecious; male flowers on long, narrow, pollen-producing catkins; female flowers in woody, rounded structures resembling small pinecones; bloom March to May; buds are stalked.
>
> *Fruit:* Small, pineconelike structures; 0.5–0.75″ long; occur on long, slender stalks; seeds float.
>
> *Bark:* Dark; smooth or rough; speckled with short, warty, horizontal stripes.
>
> *Reproduction:* By seed and spreading roots.
>
> *Control:* Cut and treat stump with a 50% a.i. solution of glyphosate.

Japanese Barberry
Berberis thunbergii

Japanese barberry is low-growing, deciduous shrub that is commonly planted as a hedge. It often escapes cultivation. Populations can be abundant in some areas, shading out other understory species. Several cultivars are available commercially. Recent research in New Jersey indicates that Japanese barberry changes the soil chemistry in environments it inhabits.

> *Habitat:* Woodland edges, open woods, roadsides, fence rows, old fields; tolerates a wide variety of soils but prefers those that are well-drained.
>
> *Height:* Typically 2–3′.
>
> *Leaves:* Simple; alternate; small; oval to spoon-shaped; smooth margins; clustered in tight bunches above spines; leaf out early in spring; may have reddish cast in spring and summer; turn attractive shades of orange, red, and purplish red in fall.
>
> *Stems:* Numerous; spiny; slightly curving; older stems are gray; twigs and young stems turn reddish brown in winter; inner bark is yellow.

Japanese barberry can form dense populations like this one along a woodland edge.

Small, yellow flowers bloom in May; spines occur individually.

Distinctive bright red berries hang on branches into winter.

Flowers: Yellow; 0.33″ wide; stalked; single or in small clusters of 2–4 blossoms; bloom in May.

Fruit: Small, bright red, egg-shaped berries; found singly or in clusters on slender stalks; mature in midsummer; remain on stems into winter; often eaten by birds and rabbits who disperse the seeds.

Roots: Shallow, but tough.

Reproduction: By seed and creeping roots; branches root freely where they touch ground.

Control: Since Japanese barberry is one of the first shrubs to leaf out in spring, they are easy to identify for control at that time. Can be pulled or dug, but must remove all connecting roots. Wear heavy gloves to help protect hands from spines. In old fields, regular mowing may be effective once large plants have been removed. Controlled burns in fire-adapted plant communities will likely control this species.

Leaves have smooth edges and are clustered in tight bunches close to branches.

Triclopyr formulated for use with penetrating oil has been used successfully for cut-stump treatment of Japanese barberry and may also be effective for basal bark treatment. Cut-stump treatment with glyphosate may also work. To minimize the amount of herbicide used for foliar spraying, plants can be cut at the base in winter or spring. Resprouts can then be sprayed with glyphosate.

Russian Olive

Elaeagnus angustifolia

Russian olive is small tree or large shrub, native to southern Europe and western Asia, that has been widely planted along roadsides because of its tolerance to salt. It is also used as a landscape plant, and has become quite invasive in the Great Plains.

Russian olives are small trees that grow 15–30′ tall.

Habitat: Grasslands, stream banks, lakeshores, roadsides, urban areas, sandy and bare mineral soils; tolerates shade.

Height and Form: 15–30′; generally rounded in shape with a loose arrangement of branches.

Leaves: Simple; alternate; narrow; lance-shaped; 1–3″ long; silvery on both sides; remain late in fall; do not change color.

Flowers: Yellowish inside, silver outside; bell-shaped; single or clustered in leaf axils; fragrant; bloom in late spring.

Twigs: Silvery scales present when young; thorns often present on ends.

Fruit: Olivelike; 0.5″ long; yellow with silvery scales; dry and mealy (unlike the fruit of autumn olive); seed remains viable in soil for three years.

Bark: Thin; comes off in elongated strips.

Reproduction: By seed, sprouting from buds on the root crown, and suckering.

Control: Girdling or cut stem and basal bark herbicide treatments will control. Metsulfuron-methyl with a surfactant is reported to be highly effective.

Characteristic silvery gray leaves are long and narrow.

Burning Bush or Winged Euonymus

Euonymus alatus

A deciduous shrub native to northeast Asia and central China, burning bush is prized by landscapers for its brilliant fall foliage. As a result many cultivars

Burning bush is widely planted for its brilliant fall color.

Elliptical leaves have fine-toothed margins.

are available, and it is widely planted. Unfortunately, it has become invasive from the eastern United States to southern Illinois.

Habitat: Prefers open woods; forests to full sun, pastures, prairies, dry to relatively moist soils, roadsides.

Height and Form: 12–20′ tall; broad with a closed crown.

Leaves: Simple; opposite; elliptical; usually less than 2–3″ long; fine-toothed margins; bright pinkish to brilliant red in fall.

Twigs and stems: Distinctive wide, corky wings.

Flowers: Inconspicuous; greenish yellow; 4 petals; bloom in May and June.

Fruit: Smooth; red in late summer.

Reproduction: By seed; often spread by birds.

Control: Cut stumps can be sprayed or painted with glyphosate.

Twigs and branches are winged and corky.

Eurasian Privets

Ligustrum obtusifolium, L. ovalifolium, L. sinense, L. vulgare

Eurasian privets are perennial shrubs that are widely planted as hedges. Once escaped from cultivation, they form dense, nearly impenetrable thickets.

Habitat: Forests and grasslands; along roadsides and other areas with disturbed soil.

Height and Form: 12–15′; upright and spreading, often as wide at it is tall.

Common privet is frequently planted as a hedge.

White flowers bloom at stem tips; leaves are opposite with smooth margins.

Black berries mature in fall.

Leaves: Simple; opposite; elliptic; 1–2.5″ long; margins are smooth; hairless; leathery and tough.

Stems and twigs: Can be leggy; buds present at ends.

Flowers: Small; white with long petals; found at ends of stems in 1–3″, cone-shaped clusters; bloom June and July.

Fruit: Shiny, small black berries; 0.33″ in diameter; borne in terminal clusters; mature September to October.

Reproduction: By seed.

Control: Small plants can be dug up. For larger plants, glyphosate can be applied to cut stems or a 2% a.i. solution of glyphosate and water plus a 0.5% nonionic surfactant can be used as a foliar spray.

White Mulberry
Morus alba

White mulberry, a deciduous tree native to China, was introduced by the British into the United States before the Revolutionary War in an unsuccessful attempt to establish a silkworm industry. Birds and mammals have since spread its seeds across the continental United States. It hybridizes with the native red mulberry (*M. rubra*). Although its ripe fruit is safe to eat, white mulberry can cause several human health problems. Its milky sap is toxic for human consumption and irritates the skin; its pollen contributes to hay fever; and its unripened fruit can cause stomach irritation, nervous system stimulation, and hallucinations.

Habitat: Open waste areas, roadsides, no-till fields, woodland edges, fencerows; tolerates poor soils, drought, air pollution, and salt; cannot grow in shade.

Height and Form: Seldom more than 15′ tall in croplands or on roadsides; can reach 20–50′ tall in other areas, sometimes 80′; has a short, thick trunk (normally 8–16″ in diameter, sometimes up to 5′); and a full, branching crown.

Leaves: Simple; alternate; minute triangular teeth on margins; broadly ovate; 2–4″ long; without lobes in older trees to irregularly lobed in younger trees; shiny, bright light green on upper surface; lower surface is pale green and smooth except for hairs along main veins (unlike red mulberry leaves, which are sandpapery above and hairy beneath); thin; contain milky sap.

Flowers: Small; greenish; without petals; clustered in dense hanging spikes; dioecious with male and female flowers usually on separate trees; male flower clusters are narrow and somewhat elongated; female flower clusters are more oval; bloom April to June.

Fruit: White to pink or purple; berrylike in tight elongated clusters; 0.5–0.75″ long; very sweet or dry and tasteless, unlike red mulberry fruits, which are red-black and tasty.

Branches: Erect or drooping with numerous narrow limbs.

White mulberry is very similar to the native red mulberry (M. rubra), *but has hairless leaves, red-brown buds, and yellow-brown bark.*

Twigs: Slender; initially reddish brown and slightly hairy; become light orange and smooth with age; exude a milky sap; several shoots produced from one node.

Winter buds: Whitish; form late in season after weather has turned cold.

Bark: Gray at first, then orange-brown or yellowish brown with shallow furrows.

Root system: Widely spreading; aggressive; may clog drains.

Reproduction: By seed.

Control: Be sure to properly identify this species before beginning control measures. Girdling is sometimes used. A basal bark treatment with triclopyr formulated for use with oil can also be effective. Imazapyr has also been effective against this species. Note: Cut stems that are buried in the soil are able to regenerate.

White Poplar
Populus alba

White poplar, a deciduous tree in the Willow family, was widely planted as a street and shade tree, but is now seldom used for those purposes. Its ability to sucker profusely can result in large colonies that shade out other vegetation. It is native to Europe, western and central Asia, and northern Africa.

Habitat: Forests, grasslands, roadsides, cultivated fields; prefers sunny conditions and moist soils.

Height and Form: 80′ tall with an irregular crown; 2–3′ diameter trunk.

Leaves: Simple; alternate; white and furry beneath; dark green above; 1–4″ wide, 2–6″ long; maplelike; irregular; long-stemmed; have 3–7 large blunt teeth or shallow lobes; tremble with slight breeze.

White poplars grow rapidly and can reach 80′ in height.

Leaves are whitish green above and hairy beneath.

Fruit: Catkins; 1.5–3″ long; appear before leaves in spring; produce pollen; males are red and gray, fluffy; females are green, fluffy with cottony seed.

Young twigs and terminal buds: Woolly.

Bark: Smooth and white; the bark of older trees is rough and dark at the base of the trunk.

Reproduction: By root sprouts, rarely by seed.

Control: Controlled burns may help eliminate young suckers, if burns are repeated for several consecutive years. Stems that are less than 2″ in diameter at breast height can

Bark is white or grayish.

be cut near ground level followed by an application of glyphosate to the cut surface. Trees that are more than 2″ in diameter at breast height can be girdled. Applying triclopyr to the cut surface may reduce potential suckering. Cut-stump treatment with glyphosate or triclopyr from June through August may also work if applied to all stems in a clone, but be prepared to control suckering. Basal bark treatment with triclopyr and penetrating oil will also help control this species.

Willows
Salix spp.

Willows are found throughout the world as both deciduous shrubs and trees. The most problematic non-native willows in the upper Midwest—white willow (*Salix alba*) and crack willow (*S. fragilis*)—are from Europe. Native willows can become overly abundant as well.

Identifying individual species is often difficult due to minute identification marks, the multitude of native and introduced willow species, and frequent hybridization. All willows grow rapidly, but several species can be aggressive in certain habitats and alter wetlands by limiting herbaceous undergrowth.

White willow is commercially available. It hybridizes readily with crack willow and weeping willow (*S. babylonica*), another non-native willow. Crack willow was introduced as a fast-growing ornamental. It escapes cultivation quite readily and often forms pure stands. Crack willow is similar to the native black willow (*S. nigra*), but its branches break off more easily.

Habitat: Moist soils, stream banks, floodplains, roadsides, yards, partially shaded and open wetlands; will not tolerate heavy shade.

Height and Form: Height and form vary widely. In terms of height, willows tend to grown taller (up to 80′) on wet sites and shorter (to 10′) on drier, less favorable sites. Willows are also shorter at northern latitudes and higher altitudes. Some willows, such as white willow and crack willow, are considered trees, while

Willows often grow in pure stands in moist soils along streambanks and in floodplains. They do not tolerate heavy shade.

many others are described as tall shrubs. Willows usually have short trunks, stout, spreading branches, and broad, irregular, open crowns.

Leaves: Simple; alternate; long and narrow; with or without teeth; leaf buds are completely covered by a single scale in winter; three tiny dots or bundle scars are within leaf scars.

> *White willow:* Green in spring, yellow in fall; dull above—the only willow with silky white hairs above and especially below; lance-shaped; 1.5–6″ inches long; 0.5–1″ wide; usually finely toothed; dense; leaf stalks are 0.125–0.25″ long and may have glands; stipules are usually absent (small if present); leaves emerge early in spring and remain late in fall.

> *Crack willow:* Green on both sides; powdery beneath; hairless when mature; lance-shaped with long-pointed tips, V-shaped bases, and saw-toothed margins; 1–7″ long; 0.5–1.5″ wide; leaf stalks 0.25–0.75″ long with glands; stipules small or lacking; buds large and somewhat sticky.

Flowers: Tiny; densely clustered on catkins; male and female flowers on separate plants; usually appear with leaves in spring.

Fruit: Small, many-seeded capsules; ripen early summer to late fall; seeds dispersed by wind and water.

> *White willow:* Seeds germinate immediately after wind or water dispersal on moist, bare mineral soil; temperature must be 70°F or above.

> *Crack willow:* Seeds germinate immediately after wind or water dispersal on moist, bare mineral soil; can remain viable in water for several days but will soon desiccate if landing on dry area; no seedbank is formed.

Twigs: Often have galls resembling pinecones.

> *White willow:* Olive-brown; nearly hairless to silky; sometimes slightly drooping; somewhat brittle at base; constantly litter the ground.

> *Crack willow:* Hairless; greenish to dark red; very brittle at base; snap off audibly, fall, and take root if soil is moist.

Bark: Thick; rough; furrowed: dark brown to gray on large trees.

Crack willow is very similar to the native black willow (S. nigra), *but the base of its leaves are tapered rather than rounded.*

Reproduction: By twigs breaking off and taking root, sprouting from roots, and seed production.

Control: Glyphosate can be used in August for cut-stump treatment or for foliar treatment of smaller plants. In wet areas, use glyphosate approved by the Environmental Protection Agency for aquatic applications. Contact your natural resources department for further control information.

Siberian Elm
Ulmus pumila

The hardiest of all elms, Siberian elm is a fast-growing, deciduous tree that can form thickets of hundreds of saplings on bare ground. It is similar in appearance to Chinese elm (*U. parvifolia*), American elm (*U. americana*), and slippery elm (*U. rubra*), although it has much smaller leaves and is not nearly as attractive as those species. Chinese elm, which is also an exotic but not as invasive, is distinctive because it flowers in late summer or fall. The tips of its leaves and the teeth are less pointed than Siberian elm. Chinese elm also has a more rounded and closed crown. Unlike Siberian elm, the leaves of native elms are usually longer, strongly asymmetrical at the base, and have paired teeth on the leaf edges. A hybrid of Siberian elm and slippery elm (*U. pumila* x. *U. rubra*) is locally established in Wisconsin.

Siberian elms are typically small trees but can reach 70' tall.

Habitat: Wet and dry soils, grasslands, roadsides, pastures; tolerates a wide range of growing conditions.

Height and Form: 50–70' feet tall; round, open crown with slender, spreading branches.

Leaves: Simple; alternate; small, rarely more than 2″ long; dark green and smooth above, paler and nearly hairless beneath; elliptical; singly toothed; tapering or rounded at nearly symmetrical bases; have soft, downy hairs when young.

Flowers: Pale green, produced in spring before leaves begin to unfold; without petals; borne in small drooping clusters of 2–5 blossoms.

Fruit: Thin, flat, waferlike; 0.5″ wide; hang in clusters; develop quickly; spread by wind; each produces one seed that is circular or egg-shaped with a smooth surface.

Leaves are rarely more than 2″ long; fruits are winged.

Greenish flowers develop before leaves unfold in spring.

Bark of mature tree.

Bark: Gray or brown; furrowed at maturity; often streaked with light-colored stains.

Reproduction: By seed.

Control: Girdling trees in late spring to midsummer is the preferred management technique. Cut-stump treatment with glyphosate or basal bark treatment with triclopyr are also effective.

European Highbush Cranberry
Viburnum opulus subsp. *opulus*

European highbush cranberry is a deciduous shrub that closely resembles the native highbush cranberry (*V. opulus* subsp. *trilobum*), but its leaf lobes are shorter and less pointed, and its fruits are bitter, causing birds to avoid them. Frequently used in landscapes, European highbush cranberry sometimes escapes into natural areas.

Habitat: Open woods, thickets, damp soils.

Height: To 15′.

Leaves: Simple; opposite; 3-lobed. Glands on the petioles are prominent and concave compared to the smaller, convex glands of the native variety.

Flowers: White; 5 petals; clustered in flat 1.5–3″ wide arrangements; outer flowers are larger than those near the center; bloom May to July.

European highbush cranberry (right) has prominent concave glands just below the leaf unlike the smaller, dome-shaped glands of the native highbush cranberry (left).

European highbush cranberry in fruit.

Fruit: Red; tart but edible, attractive to birds; seed not grooved.

Reproduction: By seed.

Control: Cut-stem treatment with glyphosate is generally effective.

VINES

Wintercreeper or Climbing Euonymus
Euonymus fortunei

Wintercreeper, or climbing euonymus, a perennial evergreen woody vine from Asia, is commonly sold as a groundcover. Cultivars have varied leaf sizes and colors. May be acceptable for use in gardens and landscapes, but should not be planted adjacent to natural areas because it spreads vegetatively, forming a groundcover that outcompetes native groundlayer species. In southern parts of the Midwest, wintercreeper has been known to overtop trees, shading them out.

Wintercreeper should never be planted as groundcover in woodlands or other natural areas.

> *Habitat:* Damp to dry forests; full sun to heavy shade; will invade relatively undisturbed habitats; tolerates poor soils and various pH levels, but does not do well in heavy wet soils.
>
> *Leaves:* Simple; opposite; paired; elliptic; typically 1–2.5″ long; finely toothed; shiny; leathery; have silvery veins; dark green, remaining green over winter.
>
> *Flowers:* Small; inconspicuous; greenish white; clustered; long-stalked; bloom in June and July.
>
> *Fruit:* Pinkish to red capsules; round; smooth; mature in autumn; most often found on climbing stems; seeds have an orange coating and are dispersed by birds, other wildlife, and flooding.
>
> *Stems and Twigs:* Narrow; spread rapidly; covered with tiny warts; have abundant aerial rootlets or trailing roots.
>
> *Reproduction:* By seed, by lateral shoots from the main branches, and by the formation of new plants that emerge from rootlets along the stem.
>
> *Control:* If wintercreeper infestations are too large to remove by digging or pulling, vines can be cut followed by cut-stem treatment of each stem with a 25% a.i. solution of glyphosate or triclopyr formulated for use with water. Treatment should be done in late autumn when most native plants are dormant or in the spring before native woodland wildflowers emerge. On severely disturbed sites, wintercreeper can be treated before spring ephemerals emerge with a foliar application of 2% a.i. solution of glyphosate, triclopyr formulated for use with water, or a mix of 2,4-D and triclopyr.

Periwinkle or Myrtle
Vinca minor

Periwinkle or myrtle, a perennial woody vine from Europe, is commonly sold as a groundcover. It may be acceptable for use in gardens and landscapes, but should not be planted adjacent to natural areas because it can aggressively outcompete native groundlayer species.

> *Habitat:* Forests, woodland edges, roadsides, moist rich soils.
>
> *Height:* 6–8″.
>
> *Leaves:* Simple; opposite; shiny; evergreen, cultivars exist that are other colors; up to 2″ long; taper at both ends; short-stemmed.
>
> *Flowers:* Blue-violet, sometimes white, depends on cultivar; 1″ wide; occur

Blue-violet flowers have white, starlike centers.

singly in leaf axils; 5 flaring petals; white starlike center; bloom in early spring, sometimes into summer and fall.

Fruit: Brown; inconspicuous; beanlike.

Stems: Long; creeping; commonly root at nodes; mat-forming.

Reproduction: By root expansion and rooting where nodes touch the ground.

Control: Can be pulled, raked up, or dug. Can also be cut or mowed in spring during its rapid growth stage followed by a foliar application of glyphosate.

FORBS

Goutweed

Aegopodium podagraria

Goutweed, an introduced plant from Europe, is an aggressive, perennial groundcover, especially on moist, well-drained soils. This member of the Parsley family may be acceptable for planting in gardens and landscapes where root expansion can be restricted (for example, between a sidewalk and house), but should not be planted adjacent to natural areas.

Habitat: Grasslands, forests, roadsides, waste places, gardens, partial sun to full shade.

Height: 4–12″.

Leaves: Ternately once or twice compound; opposite; 3–9 ovate, toothed leaflets; 3.5″ long; long leaf stems rise directly from the crown; attractive; more showy than flowers; pale green with white edges, also variegated; prone to scorch with too much sun or in drought conditions; do not change color before leaf drop in late fall.

Flowers: White; small; in dense, flat-topped clusters 3″ wide and rise 18″ high on a long stalk; bloom in June.

Fruit: Ornamentally insignificant.

Reproduction: By spreading rhizomes and seed production.

Goutweed leaves have distinctive white edges.

Control: Digging will rejuvenate the root system if all root fragments are not removed. Flower heads should be removed before seed set. Glyphosate can be applied to foliage in spring or summer.

Common Burdock
Arctium minus

Common burdock is often found in disturbed areas of bare or compacted soil. Native to Europe, common burdock is a biennial, although occasionally it may not bloom until its third or fourth year. Its leaves resemble those of rhubarb, while its Velcro-like burs are capable of ensnaring birds, bats, and butterflies.

Habitat: Roadsides, old fields, pastures, waste places, stream banks, full sun, or partial shade.

Leaves: Large; resemble rhubarb; heart-shaped basal rosette leaves with wavy margins develop during first year of growth; 12–20″ long by 12″ wide; upper surface is dull dark green and veiny; underside is pale gray green and fuzzy; hollow petioles may be purple-tinged; die back over winter; second-year leaves alternate on stem and become smaller and more ovate near the top.

Flowers: Small; red-violet; tubular; tightly clustered at ends of stems or in leaf axils; surrounded by bristly, hooked bracts that form a 0.75″-wide bur; flower heads resemble those of bull thistles; bloom July to September.

Seed heads: Thistlelike burs that stick to fur and clothing and aid in distribution; consist of many 1-seeded fruits; one plant typically produces 15,000 seeds; dispersal of burs and seeds begins in September.

Stem: Normally develops during second year, erect; many branched; hairy; grooved; hollow; grows 2–6′ tall.

Taproot: Large; thick; 1′ or more deep with branches in all directions.

Reproduction: By seed.

Control: Remove seed heads before seed set, preferably while they are still green. Young plants can be removed with a dandelion digger. Pulling large plants is difficult due to thick taproots. Plants do not tolerate frequent cultivation. To control large, first- and second-year plants before they flower, use a shovel to slice through the taproot 3–4″ below the soil surface. If it is necessary to use an herbicide, clopyralid has been used successfully to control this species. Small patches can be controlled by removing the flower head at peak bloom, then applying a strong solution of glyphosate to the cut surface.

Burdock is a biennial. Its 2–6′ flowering stalk emerges during the second year.

Large, prickly burs form at the end of the second growing season.

Large leaves develop during first year of growth.

Red-violet flowers bloom May to September.

Violet-blue bell-shaped flowers occur on one side of stem.

Creeping Bellflower
Campanula rapunculoides

Creeping bellflower, a perennial forb from Eurasia and a member of the Blue-bell family, is planted as a garden ornamental. It often escapes to natural areas where it forms patches that can crowd out native plants.

Habitat: Disturbed or rocky areas, old fields, roadsides, waste places, woods, grasslands, lawns, gardens.

Height: 1–3′.

Leaves: Simple; alternate; 1–3″ long; basal leaves are heart-shaped with long petioles and coarse, unevenly toothed margins; upper leaves lance-shaped with no or short petioles and hair beneath.

Flowers: Violet-blue to purple; drooping; numerous; scattered on one side of upper stem in leaf axils; 0.75–1.5″ long; bell-shaped with 5 points; pistils have white stalk; bloom from bottom up, June through August.

Fruit: Three-celled capsules.

Stems: Erect; unbranched; slender; contain milky juice.

Roots: Tuberlike; fibrous; white.

Reproduction: By seed and creeping underground roots.

Control: Should be pulled or cut close to the ground to prevent seed production.

Biennial Eurasian Thistles
Plumeless or bristly thistle: *Carduus acanthoides*
Nodding or musk thistle: *Carduus nutans*

These two thistles are biennial forbs in the Composite family. Plumeless thistle is classified as a noxious weed in eleven states, including Iowa and

In 1993, total direct control costs for noxious weeds were estimated between $3.6 to $5.4 billion annually, with an additional $4 billion in indirect costs. In agricultural production, invasive plants outcompete crops for soil and water resources, reduce crop quality, interfere with harvesting operations, and reduce land values. The estimated annual loss in productivity of 64 crops is $7.4 billion. (Bureau of Land Management, "Pulling Together: National Strategy for Invasive Plant Management," endorsed and supported by a collection of eighty-seven federal and state agencies and private organizations)

Plumeless thistle stems are winged and spiny; leaves deeply divided with alternate lobes; lobes end in a spine.

Nodding thistle flower heads are large and usually nod upon maturity.

Minnesota. Nodding thistle is particularly pervasive, and is classified as a noxious weed in sixteen states, including Illinois, Iowa, Minnesota, and Ohio. Though non-native thistles can be very aggressive on disturbed sites, they typically do not pose a great threat to high-quality natural areas. However, the presence of non-native thistles can lead to severe degradation of native grasslands and meadows because grazing animals focus their grazing on native vegetation, giving the thistles a competitive advantage.

Habitat: Disturbed areas, old pastures, roadsides, waste places, ditch banks; old fields, hay fields.

Height:
> *Plumeless thistle:* 3–6′.
> *Nodding thistle:* 2–7′.

Leaves:
> *Plumeless thistle:* Simple; alternate; deeply divided with alternate lobes that end with white to yellowish spines; almost hairless on upper surface; hair on lower surface more dense, especially near midrib.
> *Nodding thistle:* Simple; alternate; very spiny; dark green with light green midrib; smooth and hairless on both sides; slightly wavy; margins are coarsely lobed with each lobe ending in a prominent spine; first year basal rosette leaves can be 12″ or more long and 5″ wide.

Flower heads:
> *Plumeless thistle:* Red to purple; erect; 0.5–1″ in diameter; single or clustered; spiny wings at base.
> *Nodding thistle:* Reddish purple; on ends of stems; 1.5–3″ in diameter; surrounded by spiny-tipped bracts; usually droops or nods when mature; comprised of hundreds of tiny individual flowers; bloom begins in May with tallest flower head and continues through August on lower branches.

Seedlings: Seedlings of both species emerge from early spring to late fall.

Stems:
> *Plumeless thistle:* Erect; branched with spiny wings; begin elongation in May of second year.
> *Nodding thistle:* Appear winged with multiple branches on the lower portion; stems and wings are spiny; stems and branches are covered with short hairs.

Roots: Both species have a single taproot.

Reproduction: Both species reproduce by seed, with each plant capable of producing up to 10,000 seeds; seeds may disperse within seven to ten days of flowering; most produced on upper branches; up to 95 percent will germinate; wind may disperse seeds over long distances; remain viable in the soil for more than ten years. The two species may hybridize with each other.

Control: Non-native thistles in buffer zones near high-quality natural areas or restoration sites should be controlled before beginning a prairie restoration. Several native thistles found in prairies are sometimes confused with non-native species. Verify identification before initiating control work. Maintaining a healthy stand of native plants will help prevent weed seedlings from getting established.

For effective control of biennial thistles, the primary goal is to eliminate seed production. Close mowing or cutting of second-year plants twice per growing season just before flowering will usually prevent seed production. It is best to cut thistles at the early bud stage and again when resprouts reach the early bud stage. Mowing after flowering begins may spread viable seeds. Cutting plants with a sharp shovel 1–2″ below the soil surface before flowering will also provide effective control.

Herbicides are not recommended for use on non-native thistles in high-quality natural areas. If their use is unavoidable, herbicides are most effective when thistles are in the rosette stage and least effective when they are in flower. Spot-spraying rosettes in the fall after most native plants have gone dormant with 2,4-D ester can limit herbicide contact with nontarget plants. On sites with extensive thistle populations, 2,4-D ester can be applied 10–14 days before bolting using a backpack or tractor-mounted sprayer.

Clopyralid and metsulfuron-methyl are also used as foliar sprays to control biennial thistles. For small infestations, flower heads can be cut and bagged at peak bloom followed by a strong application of glyphosate to the cut stem surface.

Celandine
Chelidonium majus

Celandine is a biennial forb in the Poppy family. Native to Eurasia, it is available commercially in the United States. It often escapes cultivation and spreads into natural areas. It should not be confused with lesser celandine (*Ranunculus ficaria*), which is a member of the Buttercup family and is also an invasive, introduced plant.

Leaves are large and deeply lobed.

Yellow flowers have 4 petals and conspicuous stamens.

Habitat: Rich damp soils, forest edges, roadsides, waste places, urban areas.
Height: 1–3′.
Leaves: Pinnately compound; alternate; 4–8″ long; light green; divided into irregularly lobed leaflets.
Flowers: Yellow; four petals; 0.67″ wide; in small clusters; stamens conspicuous; primary bloom period is in spring with lesser bloom continuing to August.
Fruit: Slender, smooth pods; up to 2″ long.
Stems: Erect; branched; ribbed; hairy; produce an orange-yellow juice that can stain and irritate the skin.
Reproduction: By seed.
Control: Can be pulled or dug.

Oxeye Daisy

Chrysanthemum leucanthemum, syn. *Leucanthemum vulgare*

Oxeye daisy, a perennial forb in the Composite family, is native to the meadows and grasslands of Europe. It is not a threat to established prairies and savannas, but can form dense patches in disturbed areas and competes with native plants. Farmers do not like oxeye daisy because cows that eat it can produce off-flavored milk. It is listed as secondary noxious weed in Minnesota, and a variety of oxeye daisy—Lamotte oxeye daisy (*Chrysanthemum leucanthemum* var. *pinnatifidum*)—is listed as a prohibited noxious weed in Ohio. Oxeye daisy is often included in wildflower seed mixes.

Habitat: Disturbed areas, pastures, fields, roadsides, open woods, waste places.
Height: 8–30″.
Leaves: Simple; alternate; narrow; dark green; glossy; toothed or deeply lobed, especially near base of plant; basal leaves are spoon-shaped, have petioles and are 2–6″ long; upper leaves become smaller and narrower near the tops of stems and lack petioles.

Flowers: Flower heads consist of 15–30 white rays that surround a compact yellow disk with a depressed center; rays are less than 0.5″ wide; heads are 1–2″ wide and usually occur singly at the ends of stems and branches; disk flowers bloom from June to August.

Flower heads have white rays and a yellow disk that is depressed in the middle.

Stems: Slender; erect; usually unbranched; many emerge from the root crown or singly from an upturned rhizome.
Root system: Comprised of shallow, unbranched roots and rhizomes.
Reproduction: By seed and spreading rhizomes.
Control: Roots are shallow and plants are often easy to pull when soil is moist.

Lily-of-the-Valley

Convallaria majalis

Lily-of-the-valley is a perennial forb from Eurasia that is often planted as a groundcover. While acceptable in formal gardens, it should not be planted near natural areas because it tends to spread slowly but steadily by

White flower bells nod; leaves are large, fleshy, and oval-shaped.

means of its root system. Plant parts are poisonous, the poisonous ingredient being convallarin, thus its genus name.

Habitat: Forests, thickets, meadows, yards.
Height: 4–10″.
Leaves: Basal; 2 or 3; entire; oval; 6″ long; dark green.
Flowers: Small, white, nodding bells in a one-sided raceme; bloom late spring.
Berries: Pale red; small.
Reproduction: Spreads by root system.
Control: Can be dug when soil is moist.

Queen Anne's Lace or Wild Carrot
Daucus carota

A biennial and member of the Parsley family, Queen Anne's Lace is related to the cultivated carrot. In the eighteenth century, English royalty used it as "living lace," hence its common name. Originally from Eurasia, Queen Anne's Lace is listed as a noxious weed in Iowa, Minnesota, and Ohio. It may irritate the skin of

Flower cluster has a flat top, often purple center floret; leaves are fernlike.

some people, and it causes cows that eat large quantities of it to produce an off-tasting milk.

Habitat: Disturbed dry grasslands and fields, meadows, pastures, roadsides, waste places; prefers well-drained soils and full sun.
Height: 2–4′.
Leaves: Pinnately compound; alternate; with one to several finely divided, fernlike leaflets; leaves increase in size toward the base of stem; have carrotlike odor; basal leaves of first-year plants are 3–8″ long with long leaf stalks that encircle the stem at each node; basal rosettes remain green over winter; upper leaves of second-year plants are stalkless with white sheaths at leaf bases.
Flowers: Small; white; 5 petals; occur in 3–5″ wide flat umbels at ends of stems; a tiny, dark purple floret is often found in the center of the umbel; the primary umbel is comprised of numerous smaller umbels; 3–8 forked, leaflike bracts are located below the primary umbel; flowers bloom July to October during the second year of growth; the plant produces a succession of flowering stalks until it dies with the first hard frost in fall of the second year; once flowers mature and seed formation begins, the primary umbel curls inward resembling a bird's nest.
Seeds: Brownish with hooked spines that attach to clothing and animal fur and aid in dispersal; germinate in spring and again in fall; one plant can produce 1,000–40,000 seeds.
Stems: Erect; hairy; hollow; grooved; branched at top; may be reddish at base; usually have few leaves.
Root system: A long, slender, white taproot with fibrous secondary roots that become woody with age; carrotlike in smell and taste.
Reproduction: By seed.
Control: Can be hand-pulled or mowed before seed set in mid to late summer. Declines in number as native species become established. Generally does not survive cultivation. Young plants are susceptible to 2,4-D and triclopyr.

Broadleaf Helleborine
Epipactis helleborine

Broadleaf helleborine is a European forb in the Orchid family. After 1879, it was commonly planted in the Midwest by garden clubs. Although it has not been a significant problem in most of the region, it has become

After blooming, helleborine may grow 1–3' tall. Note the clasping, parallel-veined leaves.

Cypress spurge resembles leafy spurge but grows only 12" tall and has shorter, narrower leaves.

Cypress spurge before bloom.

Helleborine is an orchid with purple flowers or greenish flowers tinged with purple.

abundant in those areas near the Great Lakes with limestone bedrock.

Habitat: Forests with clay soils, thickets, disturbed areas.
Height: 1–3' tall.
Leaves: Simple; alternate; egg- or lance-shaped; clasp stem; strongly veined.
Flowers: Greenish with purple tinge or purple; heart-shaped lip forms a sac; arranged on a usually one-sided raceme; have oval to lance-shaped, green sepals; bloom in July and August.
Stems: Erect; one to several.
Roots: Fibrous; tightly clustered.
Reproduction: By seed.
Control: Can by hand-pulled.

Cypress Spurge
Euphorbia cyparissias

For detailed information on cypress spurge, see chapter 3 under, Leafy Spurge, Similar Species.

Creeping Charlie or Creeping Jenny
Glechoma hederacea

Creeping Charlie, a perennial herb introduced from Europe, is generally not a threat to established native plant communities except along woodland edges. It is a common pest in lawns and shady areas where it forms a dense mat that eliminates other vegetation. A member of the Mint family, it is toxic to horses when eaten in large quantities, either fresh or in hay.

Habitat: Disturbed areas, lawns, gardens, open woods, damp, shaded areas, pastures, roadsides.
Leaves: Simple; opposite; kidney-shaped; 0.5–1" in diameter; scalloped edges; deep green; reddish cast if growing in full sun; hairless; long-stemmed; veins radiate outward; have a somewhat offensive minty odor when crushed—unlike the garlic smell of first-year garlic mustard, a species with which creeping Charlie is sometimes confused.
Flowers: Small; bluish purple; 0.5–1" long; funnel-shaped with five equal teeth; 2-lipped; upper lip has two shallow lobes;

Creeping Charlie has bluish purple flowers; produces strong mint odor when crushed.

lower lip has 3 larger lobes; borne in clusters of 3–6 flowers in leaf axils on short, erect, flower stems; bloom April through June.

Fruit: Pods containing 4 nutlets.

Stems: Square; slender; sprawl over ground to form a thick, tangled mat; have numerous flowering branches; typically about 2′ long.

Root system: Shallow, fibrous roots; form at base and at almost every leaf node on stem.

Reproduction: By creeping stems and, to a lesser degree, by seed.

Control: Small patches can be carefully pulled or raked out when the soil is damp. Care must be taken to remove all roots since stems break off easily. Glyphosate may be used in areas totally overrun. All leaves should be well covered. Fall applications of fertilizer containing 2,4-D in lawn areas may be effective, but overtreatment often results since the entire lawn may not be affected. Spot applications of 2,4-D in combination with dicamba and mecoprop (MCPP) can be used when flowering or after the first hard frost in fall. (Dicamba may affect trees.) New plants of creeping Charlie may develop from the seed bank after initial control measures.

Creeping stems root wherever leaves emerge along the stem.

Baby's Breath
Gypsophila paniculata

Originally from eastern Europe and Siberia, this member of the Pink family, has an airy, delicate appearance that is prized in ornamental gardens and in cut and dried flower arrangements. A seemingly innocent plant, baby's breath is invasive in specific habitats. It is often included in wildflower mixes.

Habitat: Disturbed areas, especially dunes; also roadsides and railroad embankments; prefers sunny, slightly alkaline sands.

Height: 2–4′.

Leaves: Simple; opposite; small; narrowly lance-shaped; 1–4″ long; 0.4″ wide; hairless; have a single prominent vein; become smaller near top of plant; few present when plant is in bloom.

Flowers: White; tiny; 0.125–0.25″ wide; 5 petals; numerous; occur at ends of branches; sweet fragrance; bloom in July and August.

Seedpods: Egg-shaped to round capsule containing 2–5 seeds.

Stems: Bluish green; erect; wiry, much-branched; covered with powdery white film; hairless; number increases with age of plant.

Root system: Long, woody rootstalk; up to 12′ deep.

Reproduction: By seed (up to 14,000 per plant); spread when mature plant breaks off at base and tumbles in wind.

Control: Plants can be spot-burned with a propane torch in early spring or severed at least 4″ below the soil surface with a sharp shovel. Remove and properly dispose of plant tops. Monitor for resprouts and retreat when necessary. If using herbicides, applications of imazapic with methylated seed oil at bud to early bloom stage will result in improved control the following year. (Use of this chemical in areas of permeable soils, especially where the water table is shallow, may result in groundwater contamination.)

Baby's breath has small, white flowers and is especially invasive in dunes.

Orange day lily is a common garden flower.

Orange Daylily
Hemerocallis fulva

Orange daylily is not a true lily but a widely planted perennial hybrid of an Asian genus. It is acceptable for planting in gardens if they are not adjacent to open natural areas. Orange daylily is sometimes confused with the native wood lily (*Lilium philadelphicum*), although the latter species has whorled leaves along the length of the flower stem and spotted flowers. There are many daylily hybrids.

Habitat: Roadsides, meadows, waste places, full sun or partial shade; prefers rich soils.

Leaves: Basal; paired; lime green; 1–3′ long; narrow; smooth; swordlike; entire; slightly folded in center; shorter than the floral stems.

Flowers: Orange; bell-shaped; 3–4″ wide; unspotted; upward-facing; without fragrance; have 3 petals with wavy margins and 3 petal-like sepals; borne on round, leafless stems that rise 2–4′ high and branch at the top; each stem produces 5–9 flowers; flowers generally open for one day and bloom one at a time from June through July.

Root system: Consists of rhizomes and tuberous roots with fibrous roots growing from both.

Reproduction: By spreading rhizomes and tuberous roots; seeds are seldom fertile.

Control: In the fall, plow or dig up daylily beds, rake up and remove all plant parts. Glyphosate (20% a.i.) can be applied to cut surfaces after cutting the plant close to the ground. This method will likely require follow-up treatments.

Dame's Rocket
Hesperis matronalis

The invasion of exotic species is one of the most serious threats that parks face today, and if exotics are not actively and aggressively managed, the National Park System is at risk of losing a significant portion of its biological resources. (National Park Service Natural Resources Exotic Species Overview, http://www.nature.nps.gov/wv/exotics.htm.)

Dame's rocket colonies have a mixture of white, pink, and purple flowers. They are often mistaken for native woodland phlox, which has 5 petals, not 4.

Dame's rocket is a short-lived perennial in the Mustard family. Native to Eurasia, it is found in wildflower seed mixes and, once planted, easily escapes cultivation. It is often and easily confused with blue or woodland phlox (*Phlox divaricata*) because they bloom at roughly the same time. The native phlox has 5 flower petals, however, while dame's rocket, like all the mustards, has only 4.

Dame's rocket rosette.

Habitat: Moist and mesic woodlands, woodland edges, roadsides, fields, open areas.

Height: 2–3′.

Leaves: Simple; alternate; large; lance-shaped; toothed: form a basal rosette the first year that remains green through the following winter; a flowering stem with alternating leaves then emerges in the spring.

Flowers: White, pink, or purple; 0.75–1″ wide; showy; 4 petals; fragrant; bloom abundantly from mid-May through July; borne in large, loose clusters.

Seed pods: Long, narrow; upright; to 5″ long; contain abundant seeds.

Reproduction: By seed.

Control: Do not plant wildflower seed mixes that contain this species. Plants can be pulled or dug in early spring when the soil is moist. This may be necessary for several years to remove new plants emerging from the seed bank. If plants are pulled when in bloom, do not put them in a compost pile, as seed production may continue. In gardens, flower heads should be removed and properly disposed of once the bloom period ends to prevent seed production. Glyphosate or triclopyr can be used on large infestations. Apply to foliage on a warm day in late fall or early spring when native plants are dormant.

Flowers have 4 petals; alternate leaves that are sharply toothed and pointed.

Hawkweeds

Orange Hawkweed or Devil's Paintbrush: *Hieracium aurantiacum*
Yellow Hawkweed or King Devil: *H. caespitosum,* syn. *H. pratense*
Tall Hawkweed or Glaucous King Devil: *H. piloselloides, H. florentinum*

These hawkweeds are perennial forbs in the Composite family. They were introduced from Europe, tend to form large groups and may be allelopathic. The Roman naturalist Pliny the Elder believed that hawks ate the flowers to improve their vision, hence one of its common names. Should not be confused with native hawkweeds, such as Canada or Kalm's hawkweed (*H. kalmii*), which have alternate leaves on the stems.

Habitat: Fields, pastures, forest openings, roadsides, lawns; prefer sandy or gravelly soils and slightly acidic conditions.

A lawn/meadow dominated by orange hawkweed.

Orange hawkweed has several deep orange flowers at the end of a single stem.

Yellow hawkweed closely resembles orange hawkweed.

Height:
> *Orange hawkweed:* 6–24″.
> *Yellow hawkweed:* 1–3′.
> *Tall hawkweed:* 1–3.5′.

Leaves: Club-shaped basal rosettes; very hairy above and below.
> *Orange hawkweed:* 2–5″ long; margins are smooth or have minute teeth.
> *Yellow hawkweed:* 2–10″ long; untoothed.
> *Tall hawkweed:* 2–10″, untoothed, considerably less hairy than the other species.

Flowers: Dandelion-like; bracts below are covered with black hairs.
> *Orange hawkweed:* Deep orange-red; 0.75–1″ wide; single or in tight, flat-topped clusters at ends of stems; bloom June to September.
> *Yellow hawkweed:* Bright yellow; 0.5″ wide; two or more cluster at ends of stems; bloom May to August.
> *Tall hawkweed:* Bright yellow, 0.5″ wide; one or more cluster at ends of stems; bloom May to August.

Stems: Erect; usually leafless; milky juice produced when crushed.
> *Orange hawkweed:* Hairs on upper stem are often black and gland-tipped.
> *Yellow hawkweed:* Has black hairs.
> *Tall hawkweed:* Mostly hairless.

Root system: Rhizomes and stolons.
> *Orange hawkweed:* Most often have long rhizomes and short, stout stolons.
> *Yellow hawkweed:* Most often have long rhizomes and short, stout stolons.
> *Tall hawkweed:* Short rhizomes, no stolons.

Reproduction: Spread by rhizomes and stolons (orange and yellow hawkweed) and by wind-dispersed seeds (all three species).

Control: Can be controlled by repeated cultivation or hand-digging. Close mowing can prevent seed formation but will not prevent expansion of stolons and rhizomes. Competition from more aggressive native species may decrease density. Both clopyralid and 2,4-D have been effective in controlling orange hawkweed while in the rosette stage. A surfactant may be needed to enhance contact with hairy leaves.

Common St. John's Wort
Hypericum perforatum

A widely used herbal that was introduced from Europe, common St. John's wort is a perennial and one of many species of *Hypericum*. It is, however, the only introduced and ecologically invasive species of that group in the upper Midwest. It is easy to distinguish from the native species by the black dots on the margins of its flower petals and by its clusters of multiple flowers. Common St. John's wort is listed as a noxious weed in several western states.

Common St. John's wort has 5-petalled flowers and small leaves.

> *Habitat:* Fields, pastures, roadsides, open woods, dunes, disturbed ground; prefers sandy, dry soils.
> *Height:* Typically 1–2.5′.
> *Leaves:* Simple; opposite; small; 1–2″ long; oblong to elliptic; narrow; toothless; numerous, especially near top; lack hair and petioles; dots resembling pinpricks can be seen when leaves are held up to light.
> *Flowers:* Yellow; starlike; 5 petals with tiny black dots on edges; 0.5″ long and 0.75–1″ wide; numerous stamens; occur at ends of branches or stems in broad, branched, round-topped clusters; 25–100 flowers per cluster; bloom June to September.
> *Fruit:* Three-sectioned pods with numerous dark brown seeds; one plant can produce 100,000 seeds per year; seeds dispersed by wind, the fur of animals, or other means; viable for ten years.
> *Stems:* Usually reddish; erect; branched; smooth; woody at base.
> *Root system:* Consists of a long taproot with shallow rhizomes extending several inches from the crown; tough.
> *Reproduction:* Mostly by seeds, but rhizomes can also produce new plants.
> *Control:* Hand-pulling, digging, mowing, and burning have generally been ineffective at controlling this species. Regular tillage has been helpful in areas where this is feasible.

Yellow or Water-Flag Iris
Iris pseudacorus

Yellow flag iris, a perennial plant native to Europe and Africa, is widely sold in the United States for garden pools. It is the only yellow-flowered iris likely to be found in natural areas, mainly in freshwater wetlands. It often occurs in clumps, growing in up to 9″ of water. Yellow flag iris will tolerate high soil acidity and needs high levels of nitrogen for optimum growth.

> *Habitat:* Wetlands, shorelines, ditches, areas of shallow water.
> *Height:* 1–3′.
> *Leaves:* Basal; swordlike; up to 3′ long; flat; 0.75″ wide; without petioles; a basal leaf cluster surrounds an often shorter flower stalk.

Yellow water flag is the only yellow iris likely to be found in the wild.

Flowers: Deep yellow, showy; large; 3″ wide; two or three at tops of rounded stalks; has three deep yellow, petal-like drooping sepals with brownish speckled markings and three smaller, narrower, upright yellow flower petals above; bloom May through July.

Fruit: Oblong capsules, 2″ long.

Root system: Short roots and stout rhizomes.

Reproduction: By rhizomes.

Control: Can be dug up or treated with glyphosate approved for use in wet areas.

Butter-and-Eggs
Linaria vulgaris

A European perennial in the Snapdragon family, butter-and-eggs can form large colonies and crowd out other vegetation. This is especially the case on rangeland and prairie restoration sites. Butter-and-eggs received its common name from the flower colors, which resemble egg yolks and butter. Several western states list this species as a noxious weed. Butter-and-eggs is sometimes included in wildflower seed mixes.

Butter-and-eggs can be a problem in disturbed areas, especially those with sandy or gravelly soils.

Habitat: Dry fields, pastures, roadsides, woodland edges, disturbed areas, gravelly and sandy soils to fertile loams.

Height: 1–2′.

Leaves: Simple; alternate; gray-green; numerous; narrow; lance-shaped; 1–2.5″ long; smooth margins; hairless.

Flowers: Yellow with orange throat (2-lobed upper lip; 3-lobed lower lip with orange ridges); 1″ long; in elongated clusters of 15–20 flowers on each stem; bloom July to late September.

Seedpods: Egg-shaped capsules containing many seeds; seeds dispersed by wind and water; viable in soil up to eight years.

Stems: Multiple; erect; smooth; grow in clumps from rootstalks; rarely branched.

Reproduction: By seed and creeping horizontal roots.

Control: Do not plant wildflower mixes that contain this species. Intense cultivation, on appropriate sites, will help control this species. Removal by hand is effective for small patches. Herbicides are most effective when plants are in flower.

Moneywort
Lysimachia nummularia

A member of the Primrose Family and native to Great Britain and Europe, moneywort is a low-growing, perennial with leaves that resemble coins, hence its common name.

Habitat: Floodplain forests, swamps, wet meadows, stream borders, lawns, roadside ditches, grasslands.
Leaves: Simple; opposite, round; 0.5–1″ long; shiny; have short leaf stems; remain green most of the year.
Flowers: Yellow with small dark red dots; arise from leaf axils; wheel-shaped; 5 petals; 1″ in diameter; have slender stalks; bloom June to August.
Stems: Trailing; low-growing; smooth; 6–24″ long; branch frequently to form matlike growth.

Moneywort is a trailing herb with paired, roundish leaves and yellow flowers.

Reproduction: By seed and creeping stems that root at the nodes.
Control: Can be pulled or dug where practical. All stems, stem fragments, and roots should be removed. Controlled burns may be helpful in fire-adapted plant communities in early spring or late fall when native species are dormant.

Spearmint
Mentha spicata

Spearmint is a perennial that is most commonly grown as a culinary herb and groundcover. It resembles peppermint, another member of the Mint family introduced from Europe, but has multiple, smaller flower spikes rather than peppermint's single, large flower spike.

Habitat: Wet meadows, streambanks, full sun to partial shade; prefers rich soils; will not tolerate dry soil.
Height: 8–20″.
Leaves: Simple; opposite, paired; toothed; up to 3″ long; without stalks or nearly so; strong minty odor and taste.

Spearmint has pale violet or pink flowers on slender spikes; leaves have minty odor.

Flowers: Pale violet or pink; less than 0.25″ long; stamens protrude; found on upper leaf axils on slender interrupted spikes; bloom June to October.
Stems: Erect; square.
Reproduction: By spreading rhizomes and seeds.
Control: Dig out entire plant, including all the roots. Apply glyphosate to foliage in spring; retreat as necessary.

Garden or Woodland Forget-Me-Not
Myosotis sylvatica

Although beautiful, this short-lived Eurasian perennial can escape from gardens and become very dense, eliminating important native vegetation. Garden forget-me-not is very similar to true forget-me-not

Flowers are yellow, buttercup-like; plants emerge well before native vegetation. Photo by Jil Swearingen.

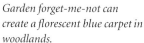

Garden forget-me-not can create a florescent blue carpet in woodlands.

Flowers have 5 blue petals and a red and yellow center.

(*M. scorpiodes*) (see Aquatic Plants) and the native bay or smaller forget-me-not (*M. laxa*). Identification should be confirmed before control measures begin.

> *Habitat:* Forests; will tolerate full sun but prefers partial shade and well-drained rich soil.
> *Height:* 6–12″.
> *Leaves:* Simple; alternate; oblong, lance or spatula-shaped; 1–3″ long; without teeth.
> *Flowers:* Florescent blue (sometimes white) with yellow eyes; small; 5 petals; showy; bloom April to September.
> *Stems:* Sprawling or weakly erect; downy; 6–20″ long.
> *Reproduction:* By abundant seed production.
> *Control:* Pull or dig out entire plant, including all the roots. Spray foliage in spring or summer with glyphosate.

Lesser Celandine
Ranunculus ficaria

Lesser celandine, a perennial in the Buttercup Family, forms thick mounds on the forest floor, excluding native ephemerals and other herbaceous plants. A European native, this species is sold commercially, and its flowers closely resemble the native marsh marigold (*Caltha palustris*) and other Buttercup family species.

> *Habitat:* Moist, forested floodplains; some upland areas.
> *Height:* 2–6″.
> *Leaves:* Arranged in low-growing rosette; kidney or heart-shaped; 1–2″ long; bluntly toothed; shiny; long-stalked; dark green; emerge well before native vegeta-

tion, sometimes by several months; aboveground portions die back by early June.
> *Flowers:* Yellow; 8–12 glossy petals; 1″ wide; borne singly on delicate long stalks; bloom in early spring.
> *Root system:* Cluster of tuberous roots.
> *Reproduction:* By seed and underground tubers.
> *Control:* Ensure proper identification before beginning control efforts. Small infestations can be pulled by hand or dug using a hand trowel or shovel. Each tuber must be removed. Glyphosate can be applied before native species come up in early spring when temperatures are 40°F or warmer.

Sheep Sorrel, Field Sorrel, or Red Sorrel
Rumex acetosella

Sheep sorrel, though somewhat attractive, can spread extensively, especially on acidic and nutrient-deficient soils. A perennial, native to Europe, it is easy to identify because of its unique, arrow-shaped leaves. This member of the Buckwheat family causes hay fever in humans and can poison livestock, if they consume sufficient quantities.

> *Habitat:* Disturbed areas, pastures, meadows, roadsides; prefers sandy or gravelly soils; does not tolerate shade.
> *Height:* 4–12″.
> *Leaves:* Simple; alternate; small, 1–3″ long; smooth; middle leaves are arrow-shaped with a pair of horizontal lobes at base; lower leaves are spade-shaped without lobes; upper leaves are small without lobes or stalks; leaves do not clasp the stem, unlike the upper leaves of garden sorrel (*R. acetosa*) or have wavy leaf margins like curled dock (*R. crispa*), both introduced sorrel species.

A prairie infested with sheep sorrel has a brownish cast in late summer.

Leaves have a characteristic arrowhead shape.

Plants are less than 12″ tall with small reddish or greenish flowers.

Flowers: Green to red or rust brown; quite small; clustered near top of plant; male and female flowers are usually on separate plants; female flowers are greenish; male flowers are yellow to red; bloom June to October.

Fruit: Seeds are reddish or golden brown; rust brown hulls often adhere to seeds; often two germination periods—one in spring and one in early fall; seeds remain viable in soil for ten to twenty years.

Stems: Upright; branched at top; slender; wiry; have sheathed nodes; several may arise from single crown.

Root system: Shallow fibrous roots and extensive horizontal roots.

Reproduction: By seeds and creeping horizontal roots that produce new shoots.

Control: Dicamba is generally effective on this species. Follow precautions on label.

Bouncing Bet or Soapwort
Saponaria officinalis

Bouncing Bet, a perennial in the Pink family, was originally planted as an ornamental, although early settlers also mixed its crushed leaves, stems, and roots with water to make a soapy lather. A European native, this species often escapes cultivation and forms dense patches. All plant parts are toxic, especially the seeds and roots.

Bouncing bet has 5 flower petals, each with indented tips.

Habitat: Disturbed areas, roadsides, fields, pastures.

Height: 1–2.5′.

Leaves: Simple; opposite, join to form a collar around the stem; 2–3″ long; 1″ wide; oval to lance-shaped; taper toward both ends; smooth margins; hairless; have 3–5 conspicuous veins on the underside; without leaf stems.

Flowers: White or pinkish; phloxlike; showy; 1″ wide; 5 petals with indentation at the tips and small appendages at flower center; petals point back from flower center; fragrant; found in dense, branched clusters at tops of stems; bloom July to September.

Seedpods: Cylindrical to oblong; contain many seeds.

Stems: Erect; usually unbranched; smooth; stout; leafy; swollen where leaves attach.

Reproduction: By seed and short rhizomes.

Control: Hand-pull small patches. Large patches can be mowed repeatedly just before flowers emerge.

White Campion, Bladder Campion, or White Cockle
Silene latifolia subsp. *alba,* syns. *S. pratensis, Lychnis alba*

This European native is a summer or winter annual or biennial, or a short-lived perennial in the Pink family. It is widespread in the northern United States and southern Canada. Easily distinguished from the native starry campion (*S. stellata*), which has fringed flower petals and whorled leaves.

Bladder campion is a biennial, blooming May to October.

Habitat: Agricultural fields, roadsides, shorelines, woodland edges, disturbed areas; prefers full sun and rich, well-drained soils.

Height: 1–4′.

Leaves: Simple; opposite; pale green; lance-shaped to oval; 1–4.5″ long; tapering to a point; margins may be wavy; downy with short hairs on upper and lower surfaces and margins; upper leaves attach directly to stem; lower leaves have long petioles; form a rosette when young.

Flowers: Showy; white (rarely pink); 1″ wide; have 5 deeply notched petals that emerge from an inflated, bladderlike structure (calyx); male and female flowers occur on separate plants; female calyx is inflated, spherical, and green with 20 lengthwise veins; male calyx is more slender, cylindrical, and purplish green with ten lengthwise veins; flowers occur in open clusters at tops of stems or singly in upper leaf axils; open at night and emit a sweet fragrance; pollinated by moths at night; bloom May to October.

Seedpods: Vase-shaped; forms inside female calyx; the calyx disintegrates after seedpod matures; mature seedpod is smooth, shiny, and light brown with an opening at the top surrounded by 10 teeth; up to 50 per plant in one season; each pod can produce 500 seeds.

Stems: Several often arise from a single taproot; erect; round; leafy; branched; covered with short hairs; have swollen nodes; may become woody with age; upper portion may be slightly sticky.

Root system: A fleshy taproot and thick lateral roots.

Reproduction: Primarily by seed; spreading rhizomes and fragmented segments of root crowns can also produce new plants.

Control: Does not tolerate cultivation. Can be difficult to control because of high seed production. Is generally susceptible to 2,4-D amine.

Tansy
Tanacetum vulgare

A member of the Composite family, tansy is often used in dried flower arrangements and sold commercially as an ornamental. Unfortunately, this European perennial frequently escapes from cultivation. Should not be confused with the rare eastern or Lake Huron tansy (*T. huronense*), which is shorter (16–32″) and has fewer, larger flowers.

Habitat: Sandy soils of open disturbed areas, roadsides, pastures, fields, prairies, hedgerows, gardens.

Height: 2–5′.

Leaves: Pinnately compound; alternate; with deeply divided, toothed fernlike leaflets; 4–10″ long, 1.5–3″ wide; smooth; strongly scented when crushed.

Flowers: Yellow, flat-topped, in buttonlike clusters; heads 0.25–0.5″ wide; numerous; showy; bloom July to October.

Stems: Erect; several arise from the base; smooth or slightly hairy; unbranched except for the flowering portion; woody; purplish red near the ground.

Reproduction: By seeds and expansion of short rhizomes. Seeds spread by wind and water.

Control: Can be dug up or cut. Mow prior to flowering; remow as needed to prevent seed set. Metsulfuron-methyl with a surfactant, clopyralid, and 2,4-D have been effective on this species.

GRASSES

Smooth Brome
Bromus inermis

Smooth brome is a perennial, cool-season grass, native to Europe and eastern Asia. It begins growth in early spring and forms a dense sod, outcompeting later growing, warm-season prairie species. Smooth brome has been widely planted as forage for livestock and erosion control along streams. Easy to distinguish from the natives, Kalm's brome (*B. kalmii*) and Canada or woodland brome (*B. pubescens*), both of which lack rhizomes and have seed heads on long, loose stems.

Habitat: Prairies, meadows, roadsides, ditches, disturbed areas; prefers well-drained fertile soils in sunny locations; will tolerate drought, extreme temperatures, and periodic flooding.

Height: Typically 2–3′.

Leaves: Rolled in the buds; blade is smooth, flat, pointed, less than 0.5″ wide, typically 4–8″ long; light green M or W-shaped marking often found between center and tip, has rough edges; sheath is round, usually smooth, edges fused except for a notch at the top; ligule is membranous

Tansy has yellow, buttonlike flower clusters.

Smooth brome inflorescence.

and short; auricles are usually absent; if present, very short and rounded.

Inflorescence: At tops of stems; 4–8″ long; comprised of 4–10 branching spikes that are 1–2″ long with 3–10 blunt-tipped florets; florets are borne in smooth, slender clusters; become purple-brown as they age; branches of flower heads are spreading when young and become erect and aligned with stem at maturity; bloom in June and July.

Seeds: Golden or tan; numerous; 0.375″ long; spread by wind, water, birds, and mammals; viable up to ten years.

Stems: Nearly smooth; erect; round.

Root system: Rhizomes.

Reproduction: By seeds and aggressive, spreading, dark-colored rhizomes.

Control: (See sidebar.)

Quackgrass

Elytrigia repens, syn. *Agropyron repens*

Quackgrass—a European sod-forming, cool-season grass—is often used for forage. A perennial, it can be very aggressive and form large patches.

Controlling Perennial Cool-Season Grasses

Many invasive, cool-season, perennial grasses (bluegrass, smooth brome, timothy, orchard grass, quackgrass, tall fescue, etc.) can be controlled using the techniques discussed below.

Manual or Mechanical Control: Prairies dominated by native warm-season grasses (e.g., big and little bluestem, Indiangrass) and forbs are typically burned in the spring to discourage non-native, cool-season grasses. Burning should ideally take place when the flowering portions of cool-season grasses are still enclosed within the sheaths. Burning at this time not only sets back non-native grasses but gives the native, warm-season grasses a competitive advantage by eliminating duff and exposing the darkened soil to sunlight. This will cause the soil to warm more quickly and encourage warm-season species to grow. Spring burning may be necessary for three or more consecutive years to achieve adequate control of non-native grasses.

Annual spring burning is not advisable in prairies with populations of native cool-season grasses and sedges since burning at this time of year will harm them. In such cases, burns may need to take place earlier in spring before native, cool-season grasses emerge, or, in eighteen-month burn cycles that alternate between spring and fall burns [Sather, TNC].

Intensive early spring grazing can also help control non-native cool season grasses. Grazing that continues throughout the growing season, however, will increase the density of non-native grasses as warm-season grasses are set back.

June or November mowing of non-native grasses will also favor the growth of native, warm-season grasses. Be aware of other desirable plant species that may be adversely affected by these methods.

To eliminate small areas of Kentucky bluegrass in yards, covering the grass for six to eight weeks during the growing season with black plastic, heavy cardboard, or other material to block sunlight has been effective. Using overlapped newspapers, ten to twelve pages thick, has also worked. Mulch can be placed on top to hold the newspapers in place. This method is often used to make planting beds out of turf if herbicide use is being avoided.

Chemical Control: On badly degraded sites that are difficult to burn because of a shortage of dry vegetation for fuel, or in urban areas where burning is prohibited, the use of a grass-specific herbicide, such as sethoxydim, in spring or fall when native grasses are dormant and the soil is unfrozen will help control non-native, cool-season grasses. If properly used, grass-specific herbicides should not harm sedges, woody plants, or broadleaf herbaceous plants.

Glyphosate with 2% a.i. can also be used on perennial, cool-season grasses when native species are dormant, but care must be used to avoid non-target species since glyphosate is non-selective.

Habitat: Open disturbed areas, roadsides, riverbanks, grasslands, pastures, croplands; tolerant of a variety of soil types; drought and salt tolerant.

Height: 2.5′.

Leaves: Rolled in the bud; blades are dull grayish green or dark green and often lighter where they meet the stem, flat, finely pointed, thin, upper surface and margins are rough or slightly hairy, lower surface is smooth, fine ribs on upper and lower surfaces, 1.5–8″ long, 0.125–0.25″ wide; sheath is round, short, margins overlap; ligule is membranous and very short; auricles are whitish green to brownish or reddish, narrow.

Inflorescence: Long; slender; unbranched spike; 2–8″ long; spikelets have 2–9 florets; blooms late May to September.

Seeds: Yellow-brown; reportedly viable for four years.

Stems: Erect; smooth; round; unbranched; with three to six joints; hollow at tip.

Root system: Mat-forming rhizomes; slender; white to pale yellow with brownish sheaths and fibrous roots at the nodes; sharply pointed tips; grows 2–6″ deep; can be 3–11′ long.

Reproduction: By seed and extensively creeping rhizomes.

Control: (See sidebar.)

Quackgrass inflorescence.

Tall Fescue

Festuca arundinacea, syn. *F. elatior*

A clump-forming, cool-season grass, tall fescue forms dense monocultures. Native to Europe, this perennial is planted for forage, erosion control, and as a drought-resistant turfgrass. Tall fescue is often infected with a fungus that suppresses the growth of other plants and makes grazing animals and small mammals that feed on it ill.

Habitat: Open areas, roadsides, grasslands, turf, grazed woodlands, waste areas, wet to dry soils; prefers cool conditions but will tolerate hot summers.

Height: 3–5′.

Leaves: Rolled in the bud; blades are dull, dark green with raised veins above and shiny below, often yellowish where they meet the stem, 2–28″ long, up to 0.5″ wide, rough margins, flat, thick, encircled by leaf sheath collar; sheath is round and smooth; ligule is short and membranous; auricles are small, round, fringed with hairs.

Inflorescence: Typically 2–12″-long panicle; may nod at tip; many-branched; branches fold against flower stem before and after flower-

Forms clumps; often planted for forage, erosion control, or as turf grass.

Tall fescue inflorescence.

ing creating a spikelike appearance; spikelets are tiny, pointed, bloom in May and June, and occur in clusters of 5–7; inflorescence becomes slightly purple as seeds mature.

Seeds: 0.25″ long; narrow; dark with purplish tinge.

Stems: Erect; smooth; round; hollow; often in clumps that produce 1–30 flowering stalks.

Root system: Tough, course roots can be very deep in moist soils; produces short, mat-forming rhizomes.

Reproduction: By seeds that mature in late summer, new shoot production from the stem base, and spreading rhizomes.

Control: (See sidebar.)

Bluegrasses
Canada bluegrass: *Poa compressa*
Kentucky bluegrass: *Poa pratensis*

Canada and Kentucky bluegrass are perennial sod-forming, cool-season grasses. Kentucky bluegrass is widely planted for turf and forage, and is often very abundant on native and restored prairies. It is more common and believed to be more aggressive than Canada bluegrass. Researchers and land managers often lump the two species together with information they present, so finding information exclusively about Canada bluegrass can be difficult. Unlike Kentucky bluegrass, Canada bluegrass is not sold commercially. Canada bluegrass is native to Europe, while there is disagreement as to whether Kentucky bluegrass is native to Eurasia, the northern United States, and Canada, or is just a Eurasian species. Because they begin growth early in the season when other

Canada bluegrass has a flat stem.

Kentucky bluegrass has a round stem.

Kentucky Bluegrass

Lorrie Otto, known internationally for her leadership in the natural landscaping movement, considers the traditional suburban lawn "immoral." In a 1981 interview, she said, "What happens in a society when the young are not stimulated by the diversity of life? Since childhood, we've been taught that one form of life depends upon another. In adulthood, we, in turn, preach it to the young. Yet in the areas where we could put our learning and teaching into practice—schoolyards, churches, hospitals, roadsides, and most obvious of all, our own yards—we neaten and bleaken, consistently and relentlessly destroying habitat for almost all life. It's as if we took off our heads, hung them up, and left them at the nature center."

species are still dormant, bluegrasses can spread quickly without competition. If not removed, they can outcompete native prairie grasses and forbs and dominate shaded areas resulting from invasion by woody species.

Habitat:

Kentucky bluegrass: Lawns, disturbed areas, pastures; roadsides, wet meadows, moist prairies, most grasslands, semi-open woods; prefers moist, calcareous soils; avoids acidic and sandy soils.

Canada bluegrass: Roadsides, meadows, waste areas, dry, sterile acidic soils.

Height: 2–16″.

Blades: Narrow; flat to V-shaped in cross section, tips shaped like the bow of a boat.

Kentucky bluegrass: Less than 0.5″ wide and 2–16″ long; usually smooth; veins are distinct; prominent midvein found beneath.

Canada bluegrass: Pale, dark blue-green.

Inflorescence: 2–6 flowering panicles.

Kentucky bluegrass: 2–6″ long panicles; branches several times at each node; rebranch to spikelets; spikelets are 0.25″ long with 3–5 florets; usually green with purplish tinge; bloom in early summer.

Canada bluegrass: 1–3″ long panicles; branches are usually paired.

Stems:

Kentucky bluegrass: Erect; rounded; 30″ high.

Canada bluegrass: Flat; broad; pointing upward; 15″ high.

Root systems: Long, creeping, densely matted rhizomes.

Kentucky bluegrass: Rhizomes often originate above ground and turn downward.

Canada bluegrass: Rhizomes almost always originate below the soil surface.

Reproduction: By seed and shallow-rooted, creeping rhizomes.

Control: (See sidebar.)

AQUATIC PLANTS

Common or Water Forget-Me-Not

Myosotis scorpioides, syn. M. palustris

Common or water forget-me-not is a perennial forb in the Forget-Me-Not family. Native to Europe, it is similar to garden or woodland forget-me-not (*M. sylvatica*) (see page 113) and the native smaller or bay forget-me-not (*M. laxa*). Its coiled flower cluster resemble a scorpion when in bud, hence its scientific name. Identification prior to beginning control efforts is essential because it is similar to other forget-me-not species.

Habitat: Shallow water, stream borders, wet soil.

Height: 4–24″.

Leaves: Simple; alternate: oblong or lance-shaped; 1–2″ long; hairy; entire; mostly without stems.

Flowers: Blue with yellow centers; 0.25–0.375″ wide; five petals; occur on one-sided racemes of two small diverging branches that uncoil as flowers begin to bloom; bloom May to October.

Stems: Sprawling; 6–22″ long; angled; slightly hairy.

Common forget-me-not has simple, alternate leaves; small, 5-petalled flowers.

Common forget-me-not grows in wet areas; resembles other varieties of forget-me-not.

Roots: Fibrous.
Reproduction: By seeds and rhizomes.
Control: Removal by hand, several time a year.

Watercress

Rorippa nasturtium-aquaticum, syn. *Nasturtium officinale*

A perennial aquatic forb that is native to Eurasia, watercress forms dense mats in slow-moving streams, springs, and other shallow, wet areas. This member of the Mustard Family has been used medicinally over the years and is currently used in salads, soups, as a garnish, and in other foods because of its sharp, peppery taste. Watercress resembles bittercress (*Cardamine* spp.), which is more erect, lacks roots along stem, and seldom forms large beds.

Watercress plants float on the water or grow in moist soils.

Habitat: Cold, clear, shallow, running water, springs, streams, sunny locations, soft sediment to gravel.
Height: 4–18″.
Leaves: Pinnately compound; alternate; up to 6″ long; divided into 3–9 oval, nearly entire leaflets; end leaflet almost round and often larger than others; partly or completely submerged; may remain green over winter in certain conditions.
Flowers: White; small, 0.25″ wide; four petals; grow in clusters; bloom April to October.
Seedpods: Slender; up to 1″ long; curve slightly upward; contain numerous coarse seeds.

leaf axil

Compound leaves of watercress have 3–9 leaflets; terminal leaflet large and nearly round; tiny white flowers.

Stems: Spreading; float in water or creep over mud; root at several points; may be several yards long; form tangled mass with other stems.
Reproduction: By seeds and plant fragmentation.
Control: Removal by hand, several times a year.

Curly-Leaf Pondweed
Potamogeton crispus

Curly-leaf pondweed is a submergent aquatic plant native to Europe. A member of the extensive Pondweed Family, its rapid, early spring growth shades out slower-growing native plants and interferes with water-based recreational activities. It dies back by mid-July.

> *Habitat:* Fresh and brackish streams, ponds and shallow lakes or lake edges to 12′ deep; grows below the water surface.
> *Leaves:* Simple; alternate; linear, oblong; 2–3″ long; wavy edges with minute teeth; upper leaves often appear waxy.
> *Flowers:* On short spikes that rise above water for wind pollination.
> *Stems:* 4-angled.
> *Root system:* Thin-branched rhizomes.
> *Reproduction:* By spreading rhizomes and fragmentation; migrating water birds may increase distribution.
> *Control:* If attempting manual removal, avoid fragmenting the plant. Herbicides, such as copper chelate, endothall and diquat, are sometimes used to control this plant. A permit will likely be needed to use herbicides in aquatic areas. Consult with an aquatic plant control expert in your natural resources department before proceeding.

Curly-leaf pondweed leaves alternate on stems; have wavy edges.

Potential Problem Species

THE SPECIES DISCUSSED in this chapter fall under one of the following categories:

- They are invasive in other areas having similar latitude or climatic conditions to the upper Midwest and are likely to pose a significant future threat to natural areas and restored sites here.
- They appear to be aggressive and are locally abundant in some areas of the upper Midwest, but it is unclear, at this time, how invasive they will eventually become throughout the region.

Land managers should watch for these species and take steps to control them in natural areas and restoration sites before they become a serious problem. Control recommendations may be unavailable or inconclusive for some species. For additional information, see chapter 2.

TREES AND SHRUBS

Siberian Peashrub
Caragana arborescens

Siberian peashrub, native to Siberia and Manchuria, is a deciduous shrub or small tree in the Legume family. It is often planted as a hedge or screen but has been escaping cultivation. There are several cultivars.

Habitat: Savannas, woodland edges, disturbed grasslands, well-drained soil; prefers full sun but will tolerate partial shade; highly tolerant of poor soils.
Current range: Widespread in areas of the upper Midwest.
Height: Up to 18′.
Leaves: Pinnately compound; alternate; 2–4″ long with 8–12 paired, oval leaflets; green, turning yellow in fall.
Flowers: Yellow; single; tubular; at ends of stalks that grow from leaf axils; bloom in May and June; start flowering when very young.
Fruit: 1–2″ long, flat, yellow-green pods; turn brown; smooth; sharply pointed; develop in June and July.

Siberian peashrub has pinnately compound leaves with 8–12 leaflets, and yellow flowers.

Reproduction: By seed.

Bark: Gray; young twigs, yellowish green.

Control: Small plants can be pulled or dug out; repeated burning will set it back; cut stump or basal bark herbicide treatments will also control.

Scotch Broom
Cytisus scoparius

Scotch broom, an upright shrub native to the British Isles as well as central and southern Europe, is a member of the Legume family. Although its stems were once used in making brooms, it is now often planted as an ornamental. Growing three feet tall the first year, it can be very aggressive, forming impenetrable stands that degrade rangeland, prevent forest regeneration, and create fire hazards. An individual Scotch broom rarely lives more than ten to fifteen years and most start to degenerate after six to eight years of growth.

Habitat: Open forests, roadsides, grasslands, and dry meadows; tolerates a wide range of soil conditions; does best in full sun on dry, sandy soils with a pH of 4.5–7.5.

Current range: Primarily in northwestern states and northern California spreading inward to the Great Plains; also found in eastern states from Maine to Georgia; and in Michigan and Ohio.

Height: 3–12′.

Leaves: Compound; alternate; absent late summer to early spring; have 3 leaflets; leaflets up to 0.5–1″ long, egg-shaped, dark green above, pale and hairy below.

Flowers: Bright yellow; pealike; 0.75–1″ long; occur individually in axils; bloom late March to June.

Fruit: Flat, pealike pods with fuzzy edges that are 1–1.5″ long; turn dark brown to black in late summer; contain 5–8 beanlike, 0.12″ long seeds that remain viable for sixty years; burst when ripe, scattering seeds for several yards.

Branches: Sharply angled; slender; may appear naked; greenish brown with five ridges when young, becoming smoother and tan with age; tips often die back late in growing season.

Reproduction: By seed or resprouting from root crown.

Control: Small plants can be pulled or dug out. Basal bark treatments with 25% a.i. triclopyr and 75% oil have been effective when used soon after flowering. 2,4-D can be used for foliar applications. Burning, on appropriate sites, helps deplete the seedbank but must be repeated two and four years later to be effective.

Princess Tree or Royal Paulownia
Paulownia tomentosa

Princess tree, native to East Asia and a deciduous member of the Figwort family, has been used as an ornamental and planted on plantations in the Northeast and Southeast for lumber. It is marketed as a "super tree" that grows rapidly, flowers when young, regenerates mature trees from coppicing (cutting back), and produces large numbers of seeds. Because they have similarly shaped leaves, princess tree is sometimes confused with native catalpa (*Catalpa speciosa, C. bignonioides*), but both catalpa species have long, slender beanlike seedpods.

Habitat: Disturbed areas, rocky cliffs, sandy streambanks, roadsides, forest edges; tolerates a wide range of habitats including dry, infertile, and acidic soils.

Current range: From New York and Massachusetts southeast to Oklahoma and Texas; also found in Ohio, Indiana, Illinois, Kentucky, and Missouri. It may be limited in its northern expansion by climate.

Height and Form: 30–40′ tall, rounded with large branches.

Leaves: Simple; opposite or whorled; 5–12″ long and 5–9″ wide; heart-shaped with long petioles; dark green; velvety on both surfaces when young, becoming smooth on top with age; drop off in summer and fall.

Flowers: Lavender or bluish; showy; 2″ long; tubular; borne in 8–12″ long, upright clusters that resemble foxglove; bloom before leaves are produced in spring; vanilla scent.

Fruit: Clustered, oval woody capsules; 2″ long, 0.5–1″ wide; light green in summer, tan in winter; resembles Pacman when they split open; abundant; each contains up to 2,000 winged seeds.

Twigs: Covered with numerous white lenticels.

Bark: Thin and rough with intermittent smooth areas; grayish brown.

Reproduction: By wind- and water-dispersed seeds and root sprouting.

Control: Seedlings can be pulled by hand when soil is loose. Cutting or girdling followed immediately by an application of glyphosate or triclopyr is the most effective control method. Basal bark treatment with triclopyr or a foliar application of triclopyr or glyphosate has been effective on smaller trees.

Corktrees

Amur corktree: *Phellodendron amurense*
Japanese corktree: *Phellodendron japonica*

Amur corktree is a deciduous tree and a member of the Citrus family. Native to eastern Asia, it has been introduced as a landscape plant into the United States because of its distinctive, corklike bark. Several cultivars are available. Japanese corktree is very similar in appearance to Amur corktree, but tends to be shorter in height. Only the female plants of both species have the potential to become invasive.

Habitat: Full sun; moist, well-drained soil; disturbed forests; adapts to many environmental conditions.

Current range:

Amur corktree: Massachusetts, Connecticut, and Illinois.

Japanese corktree: Massachusetts, New York, and Pennsylvania.

Height and Form:

Amur corktree: 30–45′ tall; broad with bold branches and short trunk.

Japanese corktree: 24–27′ tall; similar form to Amur corktree.

Leaves: Pinnately compound; opposite; 9–12″ long; have 5–13 leaflets, each 2.5–4.5″ long; dark green, turning yellow to bronze-yellow in fall.

Flowers: Dioecious; in upright flower clusters; individual flowers are inconspicuous in size; yellowish green; dioecious; bloom in June.

Fruit: Black; abundant; 0.33–0.5″ across; only found on female trees; remain on trees into late fall and winter, strongly scented.

Bark: Corky; furrowed and ridged; grayish brown.

Reproduction: By seed.

Control: Control methods for these species have not been formalized. Plant only male trees. Focus on reducing or stopping fruit production and spread. Prevent disturbances to forest areas where this species is likely to invade.

Sawtooth Oak

Quercus acutissima

Sawtooth oak is a fast-growing deciduous tree in the Beech family. Native to Asia, it produces large numbers of acorns at a young age (about seven years). Marketed as a shade tree and for wildlife food, sawtooth oaks are known to be invasive in parts of Michigan, Virginia, and Indiana and are believed to hybridize with native oaks.

Habitat: Prefers moist, well-drained, sandy-loam or clay-loam soil, and full sun; tolerates dry soil.

Current range: Northern Florida west to eastern Texas, northward through Missouri to New York. They may not be hardy in the northern parts of the Midwest.

Height and Form: 35–50′ tall, but can reach 70′; pyramidal when young; becoming rounded with age with drooping lower branches.

Leaves: Simple; alternate; shiny; oblong; 2–8″ long; dark green turning dull yellow or brown in fall; teeth pointed and bristled; resemble chestnut leaves but smaller; emerge in early spring, may persist into winter.

Flowers: Brown; inconspicuous; monoecious with pollen-bearing catkins and female flowers on the same tree.

Fruit: Brown acorns; 0.5–1″ long; cap has long, spreading, recurving scales that cover two-thirds of length.

Bark: Grayish brown; becoming deeply furrowed with age.

Reproduction: By seed.

Control: If small, seedlings can be pulled or leaves can be treated with glyphosate. To control large trees, girdle, use glyphosate for frill treatment, or cut and paint stump with glyphosate.

Saltcedar

Tamarix ramosissima

Saltcedar, a deciduous shrub or small tree in the Tamarisk family, is a significant problem in the arid West where it forms dense, impenetrable stands along river corridors. Besides outcompeting native vegetation, saltcedar can choke waterways, lower water tables, increase soil salinity, and promote flooding. Native from the Mediterranean to Japan and sold commercially in the United States, it provides little food or shelter for wildlife dependent on nutrient-rich, native plants.

Habitat: Floodplains, riparian areas, wetlands, and lake margins; can grow on highly saline soils but will tolerate alkaline conditions.

Current range: Most prevalent in western states; but also found in the Great Plains and southeastern states; reported in Missouri, Illinois, Ohio, and northern Minnesota.

Height: 5–30′.

Leaves: Compound; alternate; scalelike; 0.06″ long; tightly overlap along the stem, often encrusted with salt secretions.

Flowers: Pale pink to white; small individual flowers are arranged in dense plumes; bloom early spring to late fall.

Fruit: Capsules containing numerous, tiny seeds that are easily dispersed through the air and by water.

Bark: Smooth; reddish brown on young branches, brownish purple, ridged, and furrowed on older branches.

Reproduction: By root expansion, by resprouts if the top is damaged or removed, and by seeds.

Control: The most effective control involves a combination of mechanical, chemical, and biological techniques. Hand removal is recommended for saplings less than 1″ in diameter. Cut-stump herbicide treatment is frequently used to control large stands. Flooding is effective if root crowns can remain submerged for at least three months.

Vines

Fiveleaf Akebia or Chocolate Vine
Akebia quinata

Fiveleaf akebia is an Asian perennial in the mostly tropical Lardizabala family that can grow 20–40′ in one season. Introduced as an ornamental, it forms a dense cover that smothers existing ground vegetation and nearby trees and shrubs.

Habitat: Forests, grasslands; can invade nearly any kind of habitat; drought tolerant.

Current range: Northeast, Southeast, and southern Midwest, including Ohio, Indiana, Illinois, Michigan, and Kentucky.

Leaves: Palmately compound; alternate; with 5 leaflets; leaflets are oval, 1.5–3″ long, short-stemmed, purplish when they emerge, becoming bluish green at maturity and remaining bluish green into late fall. Deciduous in northern climates, but evergreen in the southern United States.

Flowers: Dull purple-brown; 3-parted; 1″ across; fragrant; appear late March to early April.

Fruit: Purple to violet flattened pods; 2–4″ long; contain tiny, black seeds; mature late September to early October.

Stems: Twining; woody; slender; round; green when young, brown at maturity.

Reproduction: Spreads primarily by vegetative means.

Control: Small infestations can be controlled by repeatedly cutting plants throughout the growing season. Plants can also be dug up, removing as much of the root as possible. Herbicides, such as glyphosate, provide the most effective control for large infestations.

Porcelain Berry
Ampelopsis brevipedunculata

Porcelain berry has grapelike leaves and produces colorful fruit clusters in the fall.

Porcelain berry, native to northeast Asia and Japan, is a deciduous, woody perennial in the Grape family. Extremely aggressive, it shades out native vegetation by forming a blanket on the ground and climbing over trees and shrubs, making them more susceptible to wind damage.

Habitat: Open, disturbed areas with full sun to partial shade, forest edges, grasslands, riverbanks or shorelines, waste places, and roadsides.

Current range: Abundant in the coastal zone between Boston and Washington; also found from New England to North Carolina, the eastern Midwest, and in at least one natural area in southeastern Wisconsin.

Leaves: Simple; alternate; usually deeply lobed; coarsely toothed; dark green; sometimes variegated; shiny beneath with delicate hairs along the veins; similar to grape.

Flowers: Tiny; greenish white or yellow; clustered; bloom in midsummer.

Fruit: Berries; 0.25″ in diameter; shiny; clustered; attractive in fall when they turn various shades of blue, purple, and green with small white and gray spots; eaten by birds, which disperse the seeds.

Tendrils: Positioned opposite leaves.

Reproduction: By seed and vegetatively from stem or root segments.

Control: Vines can be pulled from trees, then cut repeatedly as needed, especially in fall or spring, to prevent flower buds from forming. Mowing or repeated cutting will control porcelain berry but will not eradicate it. If using herbicide, plants can be cut in summer and the foliage of resprouts can then be treated in early fall with triclopyr or glyphosate. Basal bark applications with triclopyr formulated for use with penetrating oil have also been successful, but precautions must be taken not to harm other woody species.

Chinese Yam or Cinnamon Vine
Dioscorea oppositifolia syn. *Dioscorea batatas*

Distinctive aerial tubers, resembling tiny potatoes, are produced in the leaf axils.

Chinese yam, native to Asia, is an herbaceous, perennial in the Wild Yam family. Highly invasive, it forms matlike colonies that blanket and shade out nearby vegetation.

Habitat: Forests, streamsides, disturbed or undisturbed areas, roadsides, fencerows.

Current range: Widespread in eastern states from Vermont south to Florida, west to Oklahoma and Texas; also present in Illinois, Ohio, and Indiana.

Leaves: Simple; usually opposite but may alternate at upper nodes; spearlike to ovate; reddish purple at junction of petiole and blade.

Flowers: Greenish white; borne on spikes; bloom in June and July; have cinnamon fragrance.

Tubers or bulbils: Produced in leaf axils; resemble miniature potatoes; aerial; present June to September; can germinate two weeks after forming.

Stems: Twine counterclockwise; angled; can be 15′ long.

Reproduction: Primarily asexually by bulbil production.

Control: Cut stems can be treated with triclopyr, or a foliar application of triclopyr with a surfactant can be applied from July to October.

English Ivy
Hedera helix

English ivy, a woody evergreen in the Ginseng family, is often planted as a groundcover. Native to Europe, western Asia, and northern Africa, it is of little value to native wildlife. It can form dense colonies in woodlands that prevent the regeneration of trees, shrubs, and groundlayer species. Ivy-covered trees may be more susceptible to storm damage.

Leaves are dark green with bright white veination.

English ivy can form a dense carpet, shading out other species.

Habitat: Upland forests, forest edges, gardens, urban areas; tolerates deep shade.

Current range: Throughout the eastern, southern, and West Coast states; also found in southern parts of the Midwest and north into Wisconsin and Michigan; especially invasive in western Oregon and Washington.

Leaves: Simple; alternate; evergreen; thick; young plants are dark green with bright white veins and 3–5 lobes; becoming somewhat square to egg-shaped at 10 years of age or older.

Flowers: Yellow; small; clustered; produced in fall at tips of stems with enough sunlight, and on older plants.

Fruit: Dark blue to black; seeds spread by birds.

Stems: Can reach 1′ in diameter on older plants and climb to 90′; new plants form when soil contact is made; aerial rootlets assist in climbing.

Reproduction: Vegetatively and by seed.

Control: Do not grow English ivy near forested areas. Vines can be pulled and cut from trees, shrubs, and the forest floor. Resprouts can then be treated with glyphosate or with a 3–5% a.i. solution of triclopyr formulated for use with water. Use a surfactant for best results. Because English ivy is evergreen, it can be treated in winter months when native plants are dormant and temperatures are above 55°F for five days. Basal bark applications with triclopyr formulated for use with penetrating oil is also effective, but care must be taken to make sure that the herbicide is not absorbed by the host tree. Repeatedly burning plants with a blowtorch has also been used with some success.

Japanese Hop
Humulus japonicus

Japanese hop is an annual or short-lived perennial in the Hemp family. Its rapid growth rate can result in dense, almost solid, stands that outcompete native vegetation. Native to eastern Asia, this herb is used as an ornamental with variegated cultivars available. Unlike its close relative, the common hop plant (*H. lupulus*), Japanese hop does not contain compounds valued by brewers.

Habitat: Full sun to partial shade, disturbed areas, roadsides, open fields, river- or streambanks; prefers moist soils.

Current range: Found on the East Coast and west to the Dakotas and Kansas; upper midwestern states including Ohio, Indiana, Illinois, Iowa, Michigan, and Wisconsin.

Leaves: Simple; opposite; heart-shaped and palmately lobed; 5–9 lobes separated by V-shaped sinuses; 2–5″ long; toothed margins; rough; petioles are often longer than leaves; dark green.

Flowers: Greenish; male inflorescences are erect and 6–10″ long; female inflorescences are cone-shaped and 0.25–0.5″ long; borne in leaf axils; bloom August and September.

Fruit: Green hops produced by female plants; contain oval, yellowish brown seeds that are up to 0.25″ long; seeds believed to remain viable in soil for three years.

Stems: Up to 8–35′ in length; covered with rough hairs.

Reproduction: By seeds that are dispersed by wind and moving water.

Control: Plants can be pulled before seed set and removed from the area. If using herbicide, glyphosate can be used on foliage before plants flower. Cut-stem treatment with glyphosate may be appropriate between July and September.

Japanese Honeysuckle
Lonicera japonica

Japanese honeysuckle, a woody vine in the Honeysuckle family and native to eastern Asia and Japan, is often planted as an ornamental, although its sale and distribution are illegal in some states, including Illinois. It forms a dense blanket over trees, shrubs, and groundlayer species. Besides limiting sunlight and nutrient flow to other vegetation, it also causes top-heavy trees to break off in wind storms. It may not be hardy in northern parts of the upper Midwest.

Habitat: Open woods, woodland edges, thickets, roadsides, fencerows, prairies, disturbed areas.

Current range: Problematic throughout the United States, except for the northern Plains states and the Pacific Northwest; generally absent in Iowa and Minnesota.

Leaves: Simple; opposite; oval to oblong; 1.5–3″ long; have smooth margins and short petioles; begin growth in early spring and continue into late fall; semievergreen, persisting to late winter; do not unite at base like native honeysuckle vines (*L. dioica*).

Flowers: White to yellow; paired; tubular; occur in leaf axils along stems; have projecting curved stamens; fragrant; typically bloom April through June.

Fruit: Purple to black (native honeysuckle vines have red or orange fruit); paired; mature September to November; seeds dispersed by birds.

Stems: Hairy; woody; hollow; reach 30′ or more.

Root system: Dense; suckers extensively.

Reproduction: By seeds, suckering, and runners that sprout where they touch the soil.

Pairs of white to yellow tubular flowers with projecting stamens bloom in spring.

Control: Prescribed burns in spring in fire-adapted plant communities are helpful for control. Glyphosate can be applied to foliage in fall after native plants have gone dormant, but before a hard freeze.

Mile-a-Minute Vine or Asiatic Tearthumb
Polygonum perfoliatum

Mile-a-minute vine, native from India to eastern Asia, Japan, and the Philippines, is an annual trailing vine in the Buckwheat family. It grows rapidly, covering and eventually killing shrubs and other vegetation.

Leaves are triangular with barbs on the underside. Photo by Jil Swearingen.

Habitat: Open disturbed areas, woodland edges, wetlands, stream banks, uncultivated fields; prefers moist areas.

Current range: New York, New Jersey, Delaware, Pennsylvania, Virginia, Maryland, West Virginia, Ohio, Oregon, and Mississippi.

Leaves: Simple; alternate; shaped like a triangle with equal sides; light green; barbs on underside.

Flowers: White; inconspicuous; closed; emerge from saucer-shaped sheaths along the stem.

Fruit: Metallic blue, pea-sized berries; buoyant; seeds spread by water, birds, ants, and other animals.

Stems: Reddish; narrow; delicate; covered with barbs that attach to other plants; surrounded at intervals by circular, cup-shaped leafy structures.

Reproduction: By seed.

Control: Vines can be removed by hand throughout the summer. Wear thick gloves and durable clothing for protection from barbs. Let balled-up vines dehydrate for several days before disposal. Repeated mowing or trimming will prevent flowering and seed production. Glyphosate and clopyralid have been used effectively as herbicide treatments. Maintaining a sufficient cover of native vegetation will help prevent mile-a-minute vine from becoming established.

Kudzu

Pueraria montana (Pueraria montana var. *lobata,* syn. *P. lobata)*

Referred to as "the vine that ate the South," kudzu is extremely destructive and often covers large trees, utility lines, billboards, and even abandoned houses. A perennial member of the Legume family and native to China and Japan, this species now covers seven million acres in the southeastern United States. It was originally planted in the 1930s for fodder, erosion control, and shade. The state of Illinois, which had 105 known populations in 2001, has embarked on an extensive interagency program to try to eradicate this invasive plant.

Kudzu can engulf everything in its path. Photo contributed by Bill McClain, Illinois Dept. of Natural Resources.

Leaves are large with 3 leaflets.

Habitat: Forests, grasslands, wetlands; roadsides, steep embankments, fence-rows, abandoned fields, old home sites; does best in sunny locations on well-drained neutral soils; will tolerate drought, severe winters, and deep shade.

Current range: From New York and Massachusetts south to Florida, west to Texas; also found in Ohio, Indiana, Illinois, Missouri, Kansas, and Hawaii.

Leaves: Compound; alternate; comprised of 3 large, somewhat oval leaflets; center leaflet has a longer stalk; hairy when young but not when mature.

Flowers: Reddish purple; white or pink blossoms are rare; pealike; grow on 6″

Kudzu produces fragrant blooms in August.

upright spikes that emerge from leaf axils; bloom August through early September; sweet fragrance.

Fruit: Dark brown, flat seed pods; 2–3″ long; covered with long, spreading hairs; each can produce up to 9 seeds with 90 percent viability.

Stems: Older vines are woody; can be almost 3″ in diameter; have obvious growth rings; capable of growing 60′ in one season.

Roots: Starchy; tuberous taproots; can be 12′ deep.

Reproduction: By seed, root expansion, and fragmentation of vines; roots can form at each leaf or node.

Control: Clopyralid has been effective in controlling large populations. Triclopyr or glyphosate can be applied to cut stems or as foliar sprays once infestations have been reduced. Basal bark applications of triclopyr have also been helpful.

Swallow-worts

Black Swallow-wort: *Vincetoxicum nigrum* syn. *Cynanchum louiseae, C. nigrum*
Pale Swallow-wort: *V. rossicum* syn. *Cynanchum rossicum*

Black (also called "Louis") and pale (also called "European") swallow-worts are herbaceous perennials from southwestern Europe that have caused considerable ecological damage in New York and Ontario by outcompeting native wildflowers and young trees. It is feared these members of the Milkweed family may affect future monarch populations because butterflies that lay their eggs on these species rather than on native milkweed plants (*Asclepias* spp.) experience a much higher mortality rate. Pale swallow-wort is the most aggressive of the swallow-worts in New York and New England. White swallow-wort (*V. hirundinaria*) is also invasive, but is now found only in New York.

Members of the Milkweed Family, invasive swallow-worts pose a danger to native plants and to monarch butterfly populations.

Habitat: Most often in oak or mixed hardwood forests to heavily shaded woods; also found in disturbed sunny areas; prairies, savannas, open fields, and along roadsides; moist to dry soils.

Current range: Northeastern states from Maine west to southern Wisconsin and Illinois; West Coast, especially California. All three species are found in New York, with pale swallow-wort being the most abundant and found west to Michigan. Black swallow-wort is locally abundant at a few sites in southern Wisconsin.

Height: 2–6′ at maturity.

Leaves: Simple; opposite; narrow; oblong to ovate with pointed tips; 3″ long; dark, glossy green; smooth with a heavy waxy coating; leaf stems present; emit a pungent herbal smell when crushed.

Flowers: Purplish black with yellow centers; 5 petals with small hairs; less than 0.25″ in diameter; clustered in leaf axils; peak bloom in June; less abundant to mid-August; slight rotting odor.

Twining vine produces tiny, deep purple flowers.

Fruit: Floss-filled slender pods; often paired, 1.5–2.5″ long; similar to others in Milkweed family; produced by midsummer; split open from August to October releasing windborne seeds that often remain close to parent plant; can produce 2,000 seeds per square yard in full sun.

Stems: 3–6′ long; may be climbing or creeping; twine around themselves and other plants.

Root system: A massive collection of roots and rhizomes.

Reproduction: By seed, spreading rhizomes, and shoots from root crown of parent plant.

Control: Monitor for populations in late summer when plants turn golden yellow and pods become prominent. Initial control efforts should concentrate on plants in sunny areas since they will produce the most seeds. Burning plants is not effective and may improve conditions for seedling establishment. Hand-pulling roots is labor-intensive and difficult since the stem base is brittle. To prevent seed dispersal, pods should be removed before they open, and then burned. Triclopyr or glyphosate can be applied to foliage around mid-September. Use of a surfactant would help herbicides penetrate the waxy leaf coating. Cut-stem treatment with glyphosate is effective, but labor-intensive.

Forbs

Wild Chervil
Anthriscus sylvestris

Although not currently widespread in the upper Midwest, wild chervil is prolific in areas where it is found. It has spread rapidly in the Northeast in recent years. Native to Europe, this biennial or short-lived perennial in the Parsley

> The one process ongoing in the 1980s that will take millions of years to correct is the loss of genetic and species diversity by the destruction of natural habitats. This is the folly our descendants are least likely to forgive us. (E. O. Wilson, *Harvard Magazine,* January–February 1980, p. 21)

Wild chervil dominating a forest edge.

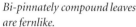

Bi-pinnately compound leaves are fernlike.

Flowers bloom in spring; bracts are absent at base of umbels.

family competes aggressively with forage crops for light, water, and nutrients and shades out surrounding vegetation. Cattle will graze on it when it is small but find it unpalatable as it matures. Sometimes confused with Queen Anne's Lace (*Daucus carota*), wild chervil blooms earlier in the season, has a more rounded umbel, and lacks the curved bracts found on Queen Anne's Lace at the base of each umbel.

Habitat: Open woods, roadsides, prairies, hay fields, pastures, waste places, disturbed areas; prefers rich, moist soils.

Current range: Maine to Virginia; the Pacific Northwest; and northern Midwest, including Michigan, Minnesota, and Wisconsin.

Height: Generally 3–4′; can reach 6′.

Leaves: Bi-pinnately or twice compound; alternate; with divided fernlike leaflets; triangular; nearly hairless; base of leaf surrounds stem; form a basal rosette the first year.

Flowers: White; small with 5-notched petals in rounded, compound umbels at tops of stems; bloom late May to early July of second year.

Fruit: Abundant; smooth; plant dies after producing seed; seeds spread by birds, water, and human activities including mowing after seed set.

Stems: Hollow; branched; hairy, especially near base.

Taproot: Up to 6′ deep.

Reproduction: By seed and lateral budding at top of roots.

Control: Pulling or digging flower stalks without removing the entire rosette and taproot will lead to resprouting the following year. Can be mowed before seed set, but root expansion will continue. Clopyralid and dicamba have been used effectively when applied shortly before blooming and one month after a prebloom cutting.

Yellow Starthistle
Centaurea solstitialis

Yellow starthistle is a winter annual in the Composite family that germinates in the fall and flowers the following May through September. Native to the Mediterranean region, it now inhabits fifteen million acres in California alone and is legally controlled in several western states where it is replacing important forage vegetation and reducing rangeland value. Livestock, wildlife, and humans find it difficult to use areas that are heavily infested with this plant. It is related to other invasive knapweed species, including spotted knapweed (*C. maculosa* syn. *C. biebersteinii*) and lesser knapweed (*C. nigra*).

Habitat: Rangeland.

Current range: Widespread; especially problematic in the West.

Height: 1.5–3′.

Leaves: In basal rosette with simple, alternating leaves on stem; become shorter and narrower with pointed tips as they ascend the stem; bases extend down the stem where they attach creating a winged appearance; dull green; covered with fine, woolly hairs.

Flower heads: Yellow with tubular florets; 1″ wide; have 0.35–0.75″ long, yellowish spines at the base in a starlike arrangement; occur singly at ends of stems.

Seeds: Large; one plant can produce 100,000 seeds; fall near parent plant.

Stems: Branched; dull green.

Root system: Taproot.

Reproduction: By seed; stick to animal fur, feathers, and clothing; human activity is primary means of dispersal.

Control: Plants can be pulled, hoed, tilled, or mowed before bloom. Several insects have been released for biological control, and various herbicides have been used to kill the plant or prevent germination. Controlled burns have been effective in California. Burns were conducted in the early flowering stage and repeated for three consecutive years. Although starthistle seeds in the soil were stimulated to germinate by the initial burn, subsequent burns helped to deplete the seed bank.

Grecian Foxglove
Digitalis lanata

Grecian foxglove, a short-lived perennial in the Figwort family from southern Europe, can create large, single species colonies. Because it is toxic to humans and animals, long sleeves and gloves should be worn when handling this plant to avoid prolonged skin contact.

Habitat: Sun to partial shade, woodland edges, roadside ditches, savannas, and prairies.

Current range: Northeast from Maine to West Virginia; upper Midwest, including Ohio, Indiana, Michigan, Wisconsin, and Minnesota; populations also present in South Carolina and Nebraska.

Height: 2–5′.

Leaves: First-year rosette leaves are dark green and spearlike; second-year leaves are simple and alternate on the stem.

Flowers: Cream-colored with purplish brown veins; tubular; conspicuous in elongated clusters; bloom in June.

Fruit: Pods with small hooks that attach easily to fur and clothing.

Stems: Coarse; erect, 1 to several.

Reproduction: By seed.

Control: Plants can be pulled or dug. Prevent seed drop. Glyphosate can be sprayed on foliage to control. Metsulfuron-methyl is also effective and more selective.

Grecian foxglove has showy, cream-colored, tubular flowers with large lips.

Hairy willow herb has widely spaced pairs of opposite leaves.

Reddish purple flower petals are notched.

Rooty projections are often observed below the stem axil.

Hairy Willow Herb
Epilobium hirsutum

Hairy willow herb, a semiaquatic perennial in the Evening Primrose family, is well established in the Northeast where it has escaped from ornamental plantings. Native to Eurasia and northern Africa, it can form dense, monotypic stands in wetlands that can crowd out cattails and other native species. It is known to compete well with purple loosestrife at some sites.

Habitat: Disturbed wet grasslands, streamsides, ditches, waste places; prefers full sun.

Current range: Pacific Northwest; Northeast from Maine to West Virginia, west to Illinois and Wisconsin.

Height: 3–6′ feet.

Leaves: Simple; opposite; usually paired; lance-shaped or oblong; sharply toothed; 2–5″ long; 0.5–1.25″ wide; without leaf stems; hairy above and below.

Flowers: Reddish purple; showy; extend from leaf axils near top of plant; 0.5–0.75″ wide; petals are notched; pistils divided into 4 sections at tip; bloom July to September; self-pollination is possible.

Seeds: Ripen and disperse 4–6 weeks after plants flower.

Stems: Erect; hairy; stout; branched.

Root system: Large; auxiliary buds at the base of stems produce stolons that root and produce branching rhizomes.

Reproduction: By rhizomes, stolons, and wind-dispersed seeds.

Control: Apply foliar herbicide by wicking or careful spraying. Bundle stems and tie, cut above string and apply herbicide to the bundle of cut stems.

Queen-of-the-Meadow or Meadowsweet
Filipendula ulmaria

Queen-of-the-meadow, native to Eurasia, is a perennial herb in the Rose family. It is sold commercially and can escape cultivation. It is sometimes confused with the native queen of the prairie (*F. rubra*), but flowers of queen of the prairie are pink while those of queen-of-the-meadow are white or cream-colored. It is unknown if this plant will become a problem.

Habitat: Moist soil, partial shade, roadsides, open damp woods.

Current range: Northeast from Maine to West Virginia; all states of the upper Midwest except Iowa and Missouri.

Height: 3–5'.

Leaves: Pinnately compound; alternate; with 5 pairs of large leaflets and many smaller ones; leaflets are elliptical, sharply toothed, light green above and white to pale green below.

Flowers: In dense white or cream-colored plumes; long stamens create fuzzy appearance; 5 petals; fragrant; bloom in summer.

Stems: Erect.

Root system: Small tubers with fibrous roots.

Reproduction: By seed.

Control: Cut or pull. Apply herbicide to cut stems if necessary.

Large, pinnately compound leaves that are green above and white to pale green below. Photo by Emmet Judziewicz.

White or cream-colored flowers help distinguish the queen-of-the-meadow from the pink-flowered, native queen-of-the-prairie. Photo by Emmet Judziewicz.

Giant Hogweed
Heracleum mantegazzianum

Giant hogweed, native to Eastern Europe and a member of the Parsley family, is a biennial or short-lived perennial. Large, painful burns and blisters will occur if its sap contacts the skin while in the presence of sunlight. Blackening of the skin and severe scarring can result. Eye exposure can lead to temporary or permanent loss of vision. Sap from leaves, and especially the stems, is most toxic when plants are in flower. If exposure occurs, wash the area immediately, stay out of sunlight, and seek medical attention. Giant hogweed, designated a noxious weed by the federal government, resembles native cow parsnip (*H. lanatum*), but is much larger. Because it grows so large so quickly, it can rapidly outcompete smaller native plants.

Habitat: Disturbed soils, roadsides, stream and riverbanks, railway embankments, fallow fields.

Current range: Northeast, Northwest, and eastern portions of the Midwest.

Height: 15–20'.

Leaves: Palmately compound with 3 deeply incised leaflets with spotted leaf stalks; enormous; lower leaves can be 5′ wide; only basal leaves produced the first year.

Flowers: White; small; many borne in very large, loose umbels at tops of stems that are up to 2.5′ in diameter; bloom late June through July.

Fruit: Flat, oval dry fruit; 0.375″ long; broadly rounded base and broad marginal ridges.

Stems: Thick; 2–4″ in diameter; hollow; dark reddish purple.

Root system: Taproot or tightly clustered fibrous roots.

Reproduction: By seed.

Control: Difficult to control because of the danger it poses to skin. Small numbers of plants can be hand-dug. Repeated mowing weakens the plants, but the large root can remain alive for many years. Foliar treatments with glyphosate or triclopyr have been effective in the Pacific Northwest. Make sure that all skin is covered when working with this plant.

Chinese Lespedeza or Sericea Bush Clover
Lespedeza cuneata, syn. *Lespedeza sericea*

Chinese lespedeza, native to eastern Asia, is a perennial in the Legume family. It was initially planted in the United States for bank stabilization, soil improvement, forage, and wildlife cover. Established stands now outcompete native vegetation in the eastern Great Plains and the Southeast. Plants become unpalatable to cattle and native wildlife after midsummer.

Habitat: Grasslands, savannas, open disturbed areas, roadsides, open woods, wetland borders.

Current range: Massachusetts and New York west to Nebraska, south to Texas, and east to Florida; found in most of the upper Midwest, except Minnesota.

Chinese lespedeza growing along a roadside.

Height: 3–5′.

Leaves: Compound; alternate; abundant; with 3 oblong, sharp-pointed leaflets, each 0.25–1″ long and 0.25–0.5″ wide; larger leaflets found on lower portion of stem; become flat at outer end; small hairs on lower surface; grayish green or silvery.

Flowers: Violet to purple-tinged; 0.25″ long; occur singly or in clusters of 2–4 in upper leaf axils; bloom late July to October; yellow when dry.

Fruit: Small; oval; one-seeded.

Stems: Erect; somewhat woody; bristly hairs on ridges.

Taproot: May be 3–4′ deep; woody; branches laterally.

Reproduction: Primarily by seeds that remain viable in the soil for twenty years or more; spread by haying and by animals after passing through their gut.

Control: Rangelands should be fertilized in April followed by burning the native grasses in late spring. Intensive grazing with mature cattle, or especially sheep and goats, should follow. Mowing in the late-flower bud stage from mid-July to late summer for two or three consecutive years will set it back. Any herbicide treatment should be completed by early to midsummer. On pastures and rangeland, metsulfuron-methyl has been effective. On noncropland areas, triclopyr or clopyralid can be used prior to branching or during flower stage but will also kill some other forb species.

Chinese lespedeza in flower.

Giant or Big-Leaf Lupine
Lupinus polyphyllus

Giant lupine is a forb in the Legume family. Although native to the Pacific Northwest and California, it is becoming locally abundant in the Northeast and Canada.

Habitat: Roadsides, grasslands, and savannas.

Current range: Most populations are found in the Northwest and Northeast; also found in the northern parts of Minnesota, Wisconsin, Michigan, and Ontario.

Height: 2–4′.

Leaves: Palmately compound; alternate; lower leaves have 12–18 leaflets.

Flowers: Blue to blue-violet; pealike; borne on 6–18″ long racemes; bloom in midsummer.

Fruit: Oblong, flattened seedpods.

Reproduction: By seed.

Control: Mow during flowering stage. Repeat as necessary. Seeds may lie dormant for many years, so monitor the site. Clopyralid can be used before seed set.

Whorled compound leaves, which are characteristic of all lupines. Photo by Robert Freckmann

In bloom, note the lengthy and large flower stalk, which is unlike the native lupine (L. perennis). Photo by Emmet Judziewicz

European Water Horehound or European Bugleweed
Lycopus europaeus

European water horehound, native to Europe, is a perennial in the Mint family. Once planted for medicinal or cooking purposes, it now lines the shores of the St. Lawrence Seaway and many other riverine areas and disturbed sites.

Habitat: Shorelines, streambanks, fields, roadsides, disturbed areas.
Current range: Northeast; also found in Alabama, Mississippi, and Louisiana, and portions of the Midwest east of the Mississippi River.
Height: 1–2.5′.
Leaves: Simple, opposite; elongated with deep, toothlike lobes; narrow; somewhat hairy.
Flowers: White with purple spots on lowest of 4 petals; small; tubular; in whorls at base of upper leaves; bristly spines on floral leaves; bloom July to October.
Fruit: Groups of individual tubular capsules with many tiny seeds.
Stems: Erect.
Reproduction: By seed.
Control: Pull or mow low prior to flowering.

Yellow Garden Loosestrife
Lysimachia vulgaris

Yellow garden loosestrife, native to Eurasia, is a perennial in the Primrose family. Planted as an ornamental, it may have the potential to become locally abundant in wetlands. It appears to be outcompeting purple loosestrife in some areas.

Habitat: Moist roadsides and woods, lakeshores, riverbanks, fens, wet meadows; full sun to partial shade.
Current range: Maine south to Virginia, west to Illinois and Wisconsin; also found in Colorado, Montana, and Washington.
Height: 2–4′.
Leaves: Simple; opposite or whorled around the stem; lance-shaped; 3–5″ long and 0.5–1.5″ wide; soft hairs beneath; dotted with black or orange glands.

Smooth leaf margins, bright yellow, 5-petalled flowers.

Flowers: Yellow with 5 petals; clustered at top of plant; 0.5–0.75″ across; lobes of calyx have reddish margins; resemble primrose; blooms from June to September.
Fruit: Dry capsules.
Stems: Erect with soft hairs.
Reproduction: Primarily by rhizomes or stolons that can be 33′ long; also by seed.
Control: Small populations can be pulled, dug, or covered with black plastic to suppress. Triclopyr or glyphosate formulated for aquatic applications may be effective.

The biological control insects used on purple loosestrife are not effective on this species.

Giant Knotweed

Polygonum sachalinensis, syn. *Fallopia sachalinensis*

(For information on giant knotweed, see chapter 3 under Japanese knotweed.)

Hedge Parsleys

Erect Hedge Parsley: *Torilis japonica*
Field or Spreading Hedge Parsley: *T. arvensis*

Erect hedge parsley and field hedge parsley are very similar in appearance. They are annual forbs in the Parsley family and originate from Eurasia.

Habitat: Forests, grasslands, hedgerows, roadsides.
Current range: Erect hedge parsley is found in the Northeast, Northwest, the South, and eastern portions of the Midwest, including Wisconsin. Field hedge parsley is more widespread and is found in southern, western, and northwestern states, and southern portions of the Midwest, including Illinois, Indiana, Ohio, and Iowa.
Height: 12–20″.

Hedge parsley (shown here) is very similar in appearance to field hedge parsley (T. arvensis), another invasive.

Leaves: Pinnately compound; alternate; sparse with 3 long, lance-shaped, toothed leaflets; center leaflet larger; prominent midrib.
Flowers: White; tiny; in loose umbels clustered at top of plant and at ends of lateral branches; up to 10 petals; bloom in July and August.
Fruit: Bristly; oval to oblong; flattened.
Stems: Erect; freely branched.
Root system: Taproot.
Reproduction: By seeds that tend to grab onto any fabric or fur that rubs against the plant.
Control: Pull or mow prior to flowering. Monitor site for additional seedlings.

Garden Heliotrope or Garden Valerian

Valeriana officinalis

Garden heliotrope, a perennial that is sometimes planted for ornamental or herbal purposes, has also been used medicinally as a tranquilizer or to help people with insomnia. Native to Eurasia, it can escape from cultivated areas and displace native vegetation in some areas.

Tiny pale pink to white flowers bloom in tight clusters.

Habitat: Open disturbed areas, grasslands, open woods, roadsides, dry to moist soils.
Current range: Northeast, Northwest, and the upper Midwest.
Height: 1–4′.
Leaves: Pinnately compound; opposite; 5–25 lance-shaped or oblong leaflets; lower leaves often have spreading, pointed teeth; petioles get smaller up the stem; underside of leaves sometimes have a few hairs.
Flowers: Pale pink or white; funnel-shaped; tiny, 0.12–0.25″ long; in tightly branched clusters; bracts are linear to lance-shaped; fragrant; bloom June to August.
Fruit: Small, lance-shaped to oblong capsules; produce abundant powdery seeds.
Stems: Stout; hairy below, especially at nodes.
Root system: Small rhizomes with fibrous roots.
Reproduction: By wind-dispersed seed and aerial stolons.
Control: Pull or mow prior to flowering. Use foliar applications of triclopyr or glyphosate.

Purpletop Verbena

Verbena bonariensis, syn. *V. patagonica*

Purpletop verbena, native to Brazil and Argentina, is an annual to short-lived perennial. Planting as a groundcover or for edging is relatively new in this area. This clump-forming species is considered invasive in the Northeast, Pacific Northwest, and parts of the South. It is uncertain if it will become invasive in the Midwest.

Habitat: Grasslands, riverbanks, roadsides, waste places, and cultivated areas; full sun to partial shade; prefers well-drained soil; heat and drought tolerant.
Current range: Eastern and western coastal states; also throughout southeastern and southern states.
Size: 4–6′ tall with a spread of 1–3′.
Leaves: Most are clustered in a mounded rosette at the base of plant; relatively scarce stem leaves are simple and opposite to nearly opposite; oval to lance-shaped; 2–5″ long; light green; narrow; sharply toothed; clasp stem; prickly to scratchy.
Flowers: Lilac purple; 0.25″ wide; borne in 2–3″ wide clusters; bloom summer to fall.
Fruit: Inconspicuous.
Stems: Upright; rigid; widely branched; square in cross-section; prickly.
Reproduction: By seed, which germinates readily.
Control: Pull new infestations. Mow before seed set.

There are currently 138 plants and 98 animals listed as endangered or threatened in Wisconsin. (Bureau of Endangered Resources, Wisconsin Dept. of Natural Resources)

GRASSES

Cheat or Downy Brome Grass
Bromus tectorum

Cheat grass, an annual from the Mediterranean region, is a major weed in winter wheat fields and western rangelands. It is highly flammable, causing a major increase in the frequency and intensity of fires on rangelands. In Idaho and Utah, more than seventeen million acres are almost totally infested. By midsummer, its stiff bristles and sharp spikelets begin to injure the eyes and mouths of livestock.

Habitat: Disturbed areas, roadsides, wastelands, range, pasture, and croplands.
Current range: Widespread throughout North America; primarily a concern in northwestern states, the Great Plains, and dry areas of the Midwest.

Highly flammable cheat grass is a serious problem in the intermontane West.

Height: To 30″.
Leaves: Blades are flat; 2–5″ long and 0.08–0.16″ wide; light green; covered with long, soft, white hairs; ligule is irregularly toothed.
Flowers: Occur in branching, very drooping, downy panicles; panicles are 3–8″ long and purplish; bloom April to May.
Spikelets: 0.75–1.5″ long; each produce 5–10 seeds.
Stems: Covered with silky hairs.
Roots: Fibrous.
Reproduction: By seed.
Control: Imazapic plus methylated seed oil can be applied before emergence or early postemergence in fall.

Japanese Stiltgrass or Nepalese Browntop
Microstegium vimineum

Japanese stiltgrass, an annual from Asia, forms dense, fast-growing mats. Although seeds are viable for only three to five years, each plant is capable of producing up to 1,000 seeds during its growing season. Native wetland and forest vegetation are particularly threatened by this species.

Habitat: Streambanks, disturbed areas, roadsides, ditches, moist forest; adapted to low light conditions.
Current range: Northeast and Mid-Atlantic areas to southern Illinois and southward to the Gulf Coast.
Height: Up to 3′.
Leaves: Lance-shaped; 2–3″ long; 0.01–0.6″ wide; alternate on stem; small hairs form a silvery stripe down center on upper surface; slightly hairy on both sides; taper at both ends; thin; pale green; slightly purplish in fall.

Japanese stiltgrass is most prolific in damp, shady areas.

Inflorescence develops in late summer or fall; a silvery stripe runs down the center of leaves.

Inflorescence: Multiple; 1–3″ long; located terminally or arise from leaf axils; spikelets are hairy and paired (one has short stalk); bloom late summer into fall.
Stems: Branched; reclining with upward pointing tips; to 3′ long; hairless.
Roots: Shallow.
Reproduction: By seed and new plant formation from rooting nodes along the stem.
Control: Hand-pulling small infestations or mowing at peak bloom in September before seeds set will help control this species. If it is necessary to use herbicides, glyphosate, an herbicidal soap such as pelargonic acid, or an herbicide specific to grasses, such as sethoxydim, are effective. In dry areas, imazapic plus methylated seed oil can be applied pre- or postemergence.

AQUATIC PLANTS

Flowering Rush
Butomus umbellatus

Flowering rush, native to Eurasia, is an emergent aquatic perennial. Introduced to the Midwest as an ornamental plant and often sold for use in water gardens. It is now illegal to buy or sell this species in Minnesota.

Plants are 1–5′ tall with rose-colored, pink, or white flower clusters.

Habitat: Muddy shores in shallow water, ditches, marshes, lakes, or streams; can grow submerged, rooted in 10′ of water and in a variety of sediments.
Current range: Northeast through northern parts of the Midwest and southern Ontario, and into the Great Plains.
Height: 1–5′.
Leaves: Swordlike; triangular in cross-section; 3′ tall; 0.5″ wide; grow from rhizomes; may remain submerged or emerge from water.
Flowers: Rose-colored, pink, or white; 0.75–1″ wide; 3 petals and 3 sepals; arranged in umbels at tops of stalks

that rise above the leaves; bloom late June to August; will not emerge or flower in deep water.

Fruit: Clustered follicles with long beaks.

Stems: Erect; triangular.

Reproduction: By seeds (often not viable in the Midwest) and short, spreading bulblet-forming rhizomes; seeds and bulblets are dispersed by water currents.

Control: Proper identification must be made before control efforts begin because when not in flower, flowering rush resembles several native shoreline plants. Plants can be cut below the water surface several times during the summer to control. Isolated plants can be dug if all root fragments are removed. Always remove plant parts from the water.

Carolina Fanwort
Cabomba caroliniana

Carolina fanwort, a perennial submersed or sometimes floating aquatic herb, is native to the subtropic-temperate regions of eastern North and South America. Often sold for use in aquariums, it sometimes finds its way into local water bodies. Carolina fanwort forms dense stands in some areas, crowding out other vegetation, clogging streams and drainage canals, and interfering with recreational, agricultural, and aesthetic water uses. Considered weedy even in its native range,

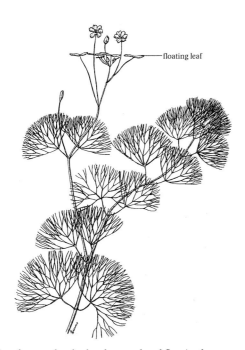

Carolina fanwort has both submerged and floating leaves.

it has created nuisance conditions as far north as New York, Michigan, and Oregon.

Habitat: Stagnant to slow-moving freshwater of lakes, ponds, rivers, and streams; water typically 3–10' deep.

Current range: Southern states including southern areas of the Midwest, north to New England, and the Northwest.

Leaves: Tightly spaced; submersed leaves have petioles and a tubular appearance, are finely divided, less than 1" long and 2" wide, fan-shaped, and arranged in pairs on the stem or whorled; floating leaves (when present) are 0.25–0.75" long, inconspicuous, diamond-shaped with stem attached in the center, and alternate on stems.

Flowers: White with yellow centers; less than 0.5" in diameter; at the end of 1–4" long peduncles; floating; bloom May to September.

Stems: Erect; 6–7' long with many branches; may be grass green, olive green, or reddish brown.

Reproduction: By short, often rooted, rhizomes and fragmentation. Reproduction by seed may be limited in northern latitudes, but shoots are capable of surviving under ice; plant parts are often spread by birds and boats; rhizomes are easily broken, facilitating vegetative spread.

Control: Research on control methods for this species is limited. Results may vary depending on the geographic region. In the South, water-level drawdowns have reduced growth in some areas but extreme drying is necessary to prevent regrowth from seed. Endothall and floridone may be effective herbicide treatments.

Water Hyacinth
Eichhornia crassipes

Native to the Amazon Basin, water hyacinth is the world's most troublesome aquatic plant. Populations of this distinctive floating plant can double in size in two weeks, shading out native aquatic plants, reducing fish populations, and displacing wildlife. Blocked waterways limit boat traffic and recreation, and may cause flooding. It is uncertain at this time if plants or viable seeds can survive midwestern winters. However, it has been observed to reproduce vegetatively and rapidly expand its population in even the northern portion of the Midwest. It is sold for use in water gardens in most states.

Habitat: Wetlands, marshes, still or slow-moving water of ponds, lakes, reservoirs, and rivers.

Distinctive air bladders act like bobbers and keep water hyacinth leaves above the water.

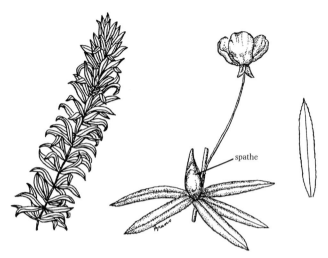

Brazilian waterweed has whorled leaves. It is often confused with hydrilla, but has larger leaves and no spines on the midrib of the leaf.

Current range: Southern, northeastern, and western states; especially problematic along the Gulf Coast and in central California; reported in Illinois and Missouri.

Leaves: Blades are cupped, round to broadly elliptic and up to 6″ wide; glossy green; have spongy stems that can be 12″ long and inflated, especially near the base; form dense, floating rosette mats that cover large areas.

Flowers: Lavender-blue with yellow blotch in upper lobe; showy; up to 2″ wide; somewhat 2-lipped; 6 petals and 6 stamens; 8–15 flowers occur on a single spike that can be 12″ long, rising above the rosette; bloom all year in mild climates.

Fruit: 3-celled capsules with many tiny seeds; form in submerged, withered flowers.

Roots: Feathery; dense near the root crown; tips have long, dark root caps.

Reproduction: By stolons that fragment and form new daughter plants, and by seed.

Control: Small populations can be removed by hand or raked up. Harvesting machines are often used to remove it in large bodies of water and canals. Glyphosate approved for aquatic applications and 2,4-D have been used to control this species.

Brazilian Waterweed
Egeria densa

Brazilian waterweed, native to Brazil, Argentina and Uruguay, is sold for use in aquariums and water gardens. It is legally regulated in some states. It can form monotypic stands that crowd out native aquatic plants. This perennial herb is found in several countries of Europe and Asia as well as Australia, New Zealand, and Chile. It diminishes fish and waterfowl, interferes with recreational activities, and supports large populations of mosquitoes.

Habitat: Rivers, streams, ponds, springs, lakes, ditches; prefers shallow, slow-moving water; can grow in depths up to 20′.

Current range: Common in much of the Southeast; also present in most Northeastern and West Coast states and in Illinois, Kentucky, Nebraska, and Kansas.

Leaves: In whorls of 4–8 around the stem; oblong to linear; 1″ long and 0.25″ wide; margins have fine teeth that require a magnifying lens to see; bright green; often confused with hydrilla (*Hydrilla verticillata*), which has smaller leaves and one or more teeth on the underside of the midrib.

Flowers: White; three petals; 0.75″ wide; situated about 1″ above the water; bloom primarily in spring to early summer but may extend into fall.

Stems: Typically 1–2′ long (can be 20′); submersed; become highly branched at water surface; bright green; are either rooted in the bottom or free-floating.

Reproduction: By fragmentation; plant parts are dispersed by birds and boating equipment; only male plants are present in the United States.

Control: Biological control, physical manipulation of habitat, and herbicides have all been used to control Brazilian waterweed.

Hydrilla
Hydrilla verticillata

Hydrilla, a submerged perennial native to Africa, Asia, and Australia, is the most troublesome aquatic plant in the United States. It tends to form single-species, mat-like stands that cover hundreds of acres. Besides out-competing native vegetation, hydrilla acts as a breeding ground for mosquitoes, and destroys fish and wildlife. Water intake and delivery systems can also be severely impacted. It is listed as federal noxious weed and is no longer sold commercially in United States. However, it is sometimes introduced as a contaminant with other aquatic plants and can be found for sale on the Internet.

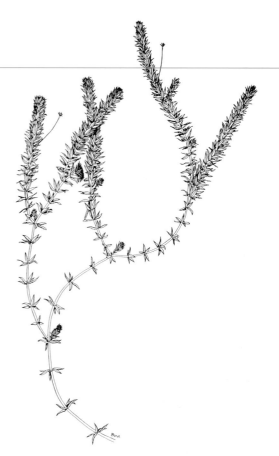

Hydrilla has whorled leaves and spines along the leaf margins and midrib.

Dense growth of hydrilla, like that shown here, makes it a serious pest in aquatic environments.

Habitat: Generally found rooted at the bottom of 20′ or more of fresh, slow-moving or still water; tolerates a wide range of growing conditions including low light, high levels of suspended sediments, drawdown periods, and warm temperatures; found in lakes, ponds, reservoirs, rivers, and ditches.

Current range: Well established in southern states; also found in California, Arizona, Washington, Virginia, Delaware, and Maryland.

Leaves: In whorls of 3–10 around the stem at nodes; 0.75″ long and 0.2″ wide; small spines give leaf margins a toothed appearance; midrib is often spiny below and reddish when new; lack stems; rough to the touch.

Flowers: Tiny; female flowers are white, located in leaf axils, have 6 petals on long, threadlike stalks; male flowers are green and look like inverted bells.

Stems: Leafy; the monoecious variety is more delicate, sprawls along lake bottoms, tends to branch at the sediment then sends up numerous stems to the surface; the dioecious variety can be up to 30′ in length and is many branched near the water surface where they spread horizontally to form dense mats.

Reproduction: Besides fragmentation, the spread of rhizomes and stolons, and some small amount of seed production, hydrilla reproduces by rhizomatous tubers and turions (long, spiny-looking buds that are produced in the fall and later break off the stem and drift or settle to the bottom where they form new plants). Tubers and turions can withstand ice cover and typical winters in the northern Midwest. Waterfowl and boats aid in dispersal.

Control: Hydrilla has similar characteristics to other aquatic plants, including Brazilian waterweed (*Egeria densa*) and the native Canadian waterweed (*Elodea canadensis*). Be sure of proper identification before beginning control measures. The aquatic herbicides, fluridone and endothall, have been used effectively for control.

European Frog-Bit

Hydrocharis morsus-ranae

European frog-bit, a perennial aquatic herb in the Frog's Bit family, is capable of forming large colonies of dense, floating mats. It can dramatically affect native aquatic life and limit recreational activities. It is often confused with American frog-bit (*Limnobium spongia*), a species whose leaves have lateral veins that make a 30–80° angle with the midvein, and whose leaf tissue contains large air pockets throughout.

Habitat: Shallow, quiet, or slow-moving water; edges of lakes, rivers, and streams; swamps, marshes, and ditches.

Current range: Present in New York, Vermont, and Michigan; also found in Quebec and Ontario and in Lake Erie and Lake Ontario.

A dense colony of European frog-bit.

Leaves: Usually floating; resemble tiny water lilies; kidney-shaped with long stems; 0.5–2.25″ in diameter; smooth; often dark purple beneath; lateral veins are arching and make a 75–90° angle with the midvein; tissue containing air pockets are located mostly along midvein.

White, 3 petalled flowers are unlike the multipetalled, native fragrant water lily (Nymphaea odorata).

Flowers: White; cup-shaped; three petals with yellow dots at base; bloom mid-summer.

Stems: Free-floating.

Reproduction: By long, cordlike stolons and vegetative buds that overwinter; seed production is limited.

Control: Other than hand pulling, there are no known control methods.

Parrot Feather Water-milfoil

Myriophyllum aquaticum, syn. *M. brasiliense*

Parrot feather water-milfoil, commonly sold for aquariums and water gardens (often under other names) is an aquatic perennial in the Milfoil family. Native to the Amazon River, its submerged form is often mistaken for Eurasian water-milfoil (see chapter 3). Parrot feather milfoil provides ideal mosquito larvae habitat and shades out algae that serve as the basis for the aquatic food chain. Infestations can hinder water movement in irrigation and drainage canals and restrict recreational opportunities.

Habitat: Quiet areas of freshwater lakes, ponds, streams, and canals; banks of muddy shores; seems to prefer nutrient-rich environments.

Current range: Has naturalized throughout southern states and northward along the East and West Coasts; also found in Ohio.

Leaves: Pinnately compound; 4–6 in numerous, feathery, bright blue-green whorls around the stem; oblong; submersed leaves are 1–1.5″ long with 20–30

A stream choked with parrot feather.

A colony of parrot feather.

Leaves are dissected like a coarse feather, thus the common name.

segments that appear to be decaying; emergent leaves are 1–2″ long with 6–18 segments; petioles are 0.25″ long.

Flowers: White; inconspicuous; small (0.06″ long); occur in emergent leaf axils in spring and sometimes in fall; male and female flowers on separate plants (only female plants are found in North America).

Stems: Stout; leafy; trail along the bottom or on the water surface; can be several yards long; erect on ends resembling little fir trees; grow up to a foot above the water surface; root at nodes; shoots begin growth early in spring and tend to die back in the fall.

Reproduction: By spreading rhizomes and fragmentation.

Control: Because parrot feather water-milfoil spreads easily by fragmentation, mechanical controls should not be used unless total takeover has occurred. Chemicals that have been used successfully against parrot feather water-milfoil include diquat, diquat and complexed copper, endothall dipotassium salt, endothall and complexed copper, and 2, 4-D. If using glyphosate formulated for aquatic use plus a surfactant, apply during the low water levels of summer and fall for more complete coverage. Consult with your natural resources department before beginning herbicide control. Biological controls may be available in the future.

Water Lettuce
Pistia stratiotes

Water lettuce, a perennial monocot native to the Amazon Basin, forms floating mats that become very dense. Plants are sold for use in aquariums and water gardens. It is uncertain if plants can overwinter in water bodies that freeze, but it can reproduce vegetatively during the summer in northeastern Wisconsin.

Habitat: Still waters of ponds, ditches, and swamps; slow-moving streams.

Current range: Southern states from North Carolina to Texas; also found in the Northeast and West.

Height: To 10″ above the water.

Leaves: A small, floating, light yellow-green, 6″-wide rosette; each leaf is 2–10″ long with distinct veins radiating from the leaf base to the leaf margin; velvety.

Flowers: White; inconspicuous; bloom in April.

Root system: Rhizomes are fleshy with long, unbranched fibrous roots that extend into the water.

Reproduction: Reproduces vegetatively.

Control: Remove by hand or raking. Copper chelate with diquat, or diquat and a surfactant, have been used to control this species.

Water Chestnut
Trapa natans

Water chestnut is a fast-growing, floating perennial herb native to southern Europe and Asia. Forming large mats that shade out native vegetation needed by waterfowl, it reduces oxygen levels for fish, encourages sedimentation by restricting silt movement, acts as a breeding ground for mosquitoes, and restricts boating and other recreational activities. Its seeds can penetrate shoe leather when stepped on, causing painful injuries. Water chestnut is illegal to sell in most southern states. (Note: This is not the species used in Asian cooking.)

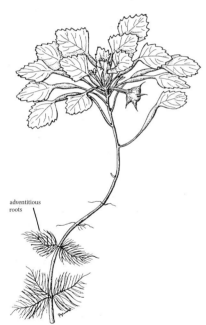

adventitious roots

Submerged leaves (also referred to as adventitious roots) are featherlike and paired, those floating on the water are waxy, triangular, and toothed.

Habitat: Full sun; typically grows in water several inches to 6′ deep; prefers water high in nutrients with a neutral or alkaline pH; can survive on muddy land exposed at low tide.

Current range: Vermont and New York, south to Maryland and Virginia.

Leaves: Submerged leaves are featherlike and paired; those floating on the surface form a rosette and are waxy, triangular, and toothed; petioles of the rosette leaves have a bladderlike swelling filled with air and spongy tissue.

Flowers: White; 4 petals; bloom mid to late July and continue until frost.

Fruit: 4.5″ long black nutlets with "horns" that have barbed spines; mature one month after flower forms; each seed can produce 10–15 rosettes, and each rosette can produce 15–20 seeds; easily dispersed by water; seeds can remain viable in the sediment for 12 years.

Stems: Cordlike; can grow 16′ long.

Reproduction: By prolific seed production and fragmentation; dispersal occurs when rosettes detach from stems and float to new areas; nutlets and seeds are spread by water currents or by attaching to the feathers of waterfowl.

Control: Can be removed mechanically or by hand. 2,4-D is the only licensed chemical known to successfully control this species.

6

Native Plants That Sometimes Need Control

On the grand scale, problems created by non-native invasive plants are many times worse than any problems caused by natives.

... The object in controlling aggressive native plants is not to eliminate them, but to promote native *plant diversity*.

PLANTS NATIVE TO THIS REGION are those that grow naturally in a particular plant community and were present before European settlement of the upper Midwest. A diverse assemblage of native plants is key to healthy natural areas, providing food and habitat for the wildlife that often rely on them to survive.

Opportunistic native plants, or colonizers, evolved to rapidly fill in disturbed areas. Disturbances can include such things as the removal of vegetation or alterations of water levels. In healthy natural areas, opportunistic native plants gradually decline in abundance and give way to slower growing native species.

Unfortunately, in today's world, with the massive and repeated disturbances to natural areas brought on by man, including the suppression of fire, many delicate or slower-growing native plants have been overwhelmed. Their numbers have declined, and their seeds may no longer be part of the soil seedbank. Native prairies and oak savannas, for example, are now extremely rare due to fire suppression, conversion to farmland, cities, and other developments. They are especially vulnerable to invasion by aggressive, non-native plants and by woody species that may be either native or non-native in origin. Therefore, instead of simply allowing nature to take its course, the survival of prairie and savanna species needs to be encouraged whenever possible to sustain biological diversity and wildlife habitat.

It must be remembered that although certain native plants may present a problem at the local level, on the grand scale the ecological problems created by non-native invasive plants are many times worse than problems caused by natives. Remember, as well, that the object in controlling native plants is not to eliminate them from all plant communities, but to promote native plant diversity whenever possible.

Control information for some native plants is not always well documented. Reviewing chapter 2, "Invasive Plant Control Techniques," and Appendix A, "Resources for Information About Invasive Plants," may provide additional helpful information when considering control for the species discussed below.

TREES

Boxelder

Acer negundo

Boxelder is a deciduous tree in the Maple family. It grows fast but has a relatively short life span. Fruits attract harmless, but sometimes annoying, boxelder beetles. Its weak wood has limited commercial value and breaks easily,

Boxelder is often found in disturbed, moist areas.

Compound, irregularly toothed leaves and winged fruits (samaras).

forming cavities or dens for squirrels, raccoons, and woodpeckers. Deer and squirrels eat the leaves. Songbirds eat the seeds, buds, and flowers.

Habitat: Prefers low, wet areas such as riverbanks, floodplains, ditches, moist disturbed areas; also found in open woods, old fields, and along fencerows and railroads.

Height and Form: 30–60′; crown is wide, open and irregular; trunks are short, can be up to 4′ in diameter; branches near the ground.

Leaves: Compound; opposite; 5–8″ long; with 3 oval, paired leaflets (sometimes 5–7) and an additional leaflet on the end; leaflets are 2–4″ long, 1–1.5″ wide, and irregularly toothed with long pointed tips; light green above, paler below; turn yellowish in fall; resemble those of poison ivy.

Flowers: Yellow-green; usually appear before leaves in spring; dioecious; male flowers are borne singly on threadlike, drooping stalks that occur in clusters; female flowers are clustered along a dangling stalk.

Female flowers in slender, hanging clusters.

Bark is grayish to light brown and grooved on older trees.

Fruit: Produced by female trees; winged; tan; 1–1.5″ long; attached to twigs in clusters of V-shaped pairs; mature September to October; often remain on tree through winter; result in many seedlings.

Stems: Branches are fragile; twigs are green or purplish.

Bark: Gray to light brown becoming dark brown and grooved with age.

Root system: Shallow; spreading.

Reproduction: By prolific seed production; will sprout from roots if injured by fire or cutting.

Control: Basal bark treatment with triclopyr formulated for use with oil is generally effective on this species. Cut-stump treatment with glyphosate or triclopyr is also effective.

Honey Locust
Gleditsia triacanthos

Honey locust, a deciduous tree in the Legume family, is a popular lawn and street tree. Many cultivars are available, including "Sunburst" locust, thornless, and few-fruited varieties. Goats, cattle, small mammals, and some birds eat the seedpods. Deer and rabbits eat the bark in winter. Provides a source of pollen and nectar for honey. Wood is durable and used for fence posts and construction. Grows from the western Appalachian Mountains west to South Dakota and north to southern Ontario.

Honey locust has bi-pinnately compound leaves and stout thorns.

Habitat: Floodplains; in moist soils of mixed forests; sometimes found in drier upland sites.

Height and Form: To 70′; open, spreading crown with a graceful, delicate silhouette; trunk may be 2–3′ in diameter.

Leaves: Bi-pinnately compound; alternate; 4–15″ long with many wavy-edged or slightly toothed leaflets; leaflets are 0.38–1.25″ long, shiny, symmetrical, fine-textured; dark green above, paler and smooth below; turn yellow in fall; produce only light shade; leafstalks are never thorny.

Thorns: Several inches long; stout; branched; clustered on trunk (sometimes surrounding it) and on stems, making identification easy.

Flowers: Yellow-green; small (0.38″ long); five petals; hang in 2–3″ clusters from the base of leaves; male and female flowers on separate trees; bloom in late spring.

Fruit: Long, flat, twisted, reddish or dark brown pods; 6–16″ long; 1.25″ wide; contain many beanlike seeds; fall to the ground unopened in late fall or winter creating a mess in urban settings; pulp is sweet and relished by some animals.

Bark: Nearly black; smooth on young trees; has long, scaly ridges when mature.
Reproduction: By seed.
Control: Basal bark treatment with triclopyr formulated for use with oil or dicamba formulated for cut-stump treatment are generally effective on this species. Fire will top-kill young plants, but it will resprout after fire or cutting.

Eastern Redcedar
Juniperus virginiana

Eastern redcedar was once relatively rare in Wisconsin and northern Minnesota, being generally confined to fire-resistant barrens and cedar glades. Due to the suppression of fire, this evergreen member of the Cypress family is now widespread, often in pure stands on lands that were once dry to mesic prairies. It can be a significant threat to native prairies, where it shades out the grasses, forbs, and other vegetation. It spreads rapidly once seed production begins, about ten years after germination. Eastern redcedar is also an alternate host for cedar-apple rust, which can affect crabapple and apple trees.

Eastern redcedar provides cover for many birds, and its berries are a staple in the diet of numerous bird species and other forms of wildlife. Deer eat its twigs and leaves. It is a popular ornamental and shelterbelt species, and the wood is used for cedar closets, chests, posts, pencils, and shingles.

Habitat: Limestone hillsides, dry gravelly soils, roadsides, old fields, hedgerows, floodplains, wetlands; tolerates temperature and moisture extremes.
Height and Form: 40–60'; narrowly upright and dense or broadly pyramidal; the trunk ranges up to 2' in diameter.
Leaves: Mostly opposite; needlelike on young shoots; scaly on older twigs; tiny (0.16 to 0.75" long); in four rows along twigs; dark green in summer; brownish green in winter; fragrant when crushed.

Eastern redcedars with prairie in the foreground.

Fruit: Small cones that look like frosty, dark blue berries; 0.25″ in diameter; on female trees; fragrant.

Twigs: 4-sided.

Bark: Shreddy; not ridged.

Wood: Red-grained with rose-brown heartwood; fragrant.

Reproduction: By seed, often dispersed by birds.

Control: Controlled burns will kill small trees if enough dry grass and other flammable vegetation is present. Cutting trunks at the base will also kill eastern redcedar. Herbicide application to cut stumps is not necessary. Winter, when there is a good layer of snow on the ground, is usually the best time to cut trees since the ground will be frozen, preventing damage to prairie vegetation from trampling and dragging to create brush piles. Brush piles should be burned; dead eastern redcedar trees can persist for decades and continue to shade out other species. Burning brush piles when there is a good layer of snow on the ground will help limit fire damage to prairie vegetation from the sustained heat.

Osage Orange
Maclura pomifera

Osage orange is a deciduous tree native to northern Texas, southeastern Oklahoma, and Arkansas. A member of the Mulberry family, it is one of the few tree species with thorns that also has a milky sap. Because of its thorns, these hardy trees were planted close together and used as a living fence before barbed wire became available. It is common in the southern part of the Midwest.

Habitat: Open sunny areas, hedgerows, fence lines, pastures, disturbed (often grazed) forests, riverbanks, prairies, savannas.

Height and Form: 20–50′; rounded; spread of crown equal to height; trunk short, 1.5–3′ in diameter.

Leaves: Simple; alternate; not toothed; 2–6″ long; shiny; oval to lance-shaped with long pointed tips; medium to dark green, turning yellow in fall.

Flowers: Tiny; male plants form short flower clusters, female plants have flowers crowded into spherical heads; bloom May and June.

Fruit: Green-yellow; wrinkled; rounded; grapefruit-sized, 4–6″ in diameter; contain many seeds that are dispersed great distances by wildlife; hazardous and messy from September to October; anyone working with fruit-bearing osage orange should wear a hardhat.

Stems: Orange-brown; zigzag; spines on vigorous young growth, may be absent on mature trees.

Sap: Milky; may cause skin rash.

Bark: Light gray-brown tinged with orange; furrowed; separates into shaggy strips on mature trees.

Reproduction: By seed and root sprouts.

Control: Cut-stump or basal bark treatments with triclopyr herbicide are effective. Periodic burning hinders the establishment of young plants. Cutting produces numerous, thorny resprouts.

Aspen, Poplar, or Popple
Big- or Large-Tooth Aspen: *Populus grandidentata*
Quaking or Trembling Aspen: *P. tremuloides*

Big-tooth aspen and quaking aspen are deciduous trees in the Willow family. Although quaking aspen is more common than big-tooth aspen, populations of both species have greatly increased due to logging and subsequent fires that have stimulated suckering. They are now widely distributed, often in dense stands with clones spread over a wide area. Prairies, barrens, and oak savannas are particularly threatened by their spread.

Aspen trees grow rapidly (exceeding 3′ per year for the first decade) and resprout easily, producing quick cover for burned-over or cleared areas. Trees reach maturity in fifty years, then decline rapidly. Rotted trees provide important cavities for wildlife. Catkins and buds provide food for birds. Deer, hares, porcupines, and beavers browse on leaves and young stems. Beaver also use trunks and branches for their dams and lodges. Aspens are an important source of paper pulp, excelsior, matches, lumber, particleboard, pallets, and boxes.

Quaking aspen along a roadside in fall. Flat leaf stems cause the leaves to tremble in the slightest breeze.

Habitat: Shade intolerant; usually found on fairly even-aged stands in forests, woodland edges or in forest gaps created by fire or tree harvesting. Can move into grasslands.

> *Big-tooth aspen:* Average to dry soils; tolerates drier conditions than quaking aspen.

> *Quaking aspen:* Most often in moist woods or along streams; grows in a wide variety of soils.

Height and Form: 20–50′ (sometimes 70′); narrow, rounded crowns, 20–30′ wide; trunks are 1–2′ feet in diameter; from a distance, the clonal stands appear dome-shaped with the tallest trees in center.

Leaves: Simple; alternate; rather sparse; nearly round to heart-shaped; slightly longer than broad with pointed tips; teeth are paired; turn gold in fall.

> *Big-tooth aspen:* 2–6″ long; large, rounded teeth on margins; white-wooly beneath when young; buds are dull and hairy; end bud is 0.38″ long; leaf stalks are present; leaves open later than quaking aspen.

> *Quaking aspen:* 1.5–3″ long; fine teeth on margins; shiny green above; dull green beneath; buds are hairless and shiny; leaf stalks are long, thin, and flat, allowing them to tremble with the slightest breeze.

Flowers: Silvery gray; dioecious; on drooping catkins that open in early spring before leaves emerge.

Fruit: Yellow-green; clustered on drooping catkins with silvery white seeds that mature in late spring and are easily dispersed by the wind; produce many seeds, but must germinate within a few days of dispersal in order to survive.

Bark: Mostly smooth; base becomes dark gray, thick, rough, and furrowed with age.

> *Big-tooth aspen:* Mostly greenish gray to yellowish.

> *Quaking aspen:* Pale greenish white, cream-colored, or chalk white.

Reproduction: Primarily by root sprouts; wind-dispersed seed.

Control: Do not plant aspens near prairies or savannas. Regular prescribed fires will kill seedlings and top-kill young resprouts, but will encourage further germination and resprouting of roots. If practical, girdling is the preferred management technique since it is effective and minimizes resprouting. Girdles should be at least 6″ wide and should be cut in the spring before leaves fully develop. For maximum effect, girdle each tree in the clone. Avoid girdling with a saw since cuts may be too deep, damage the woody inner tissue, and stimulate sprouting. Stems that are too small to be girdled can be cut twice in one year, each time after full leaf-out. If trees must be removed, wait until they are completely dead before cutting them down. A common mistake is to cut down large trees without any other treatment. This will lead to massive sprouting of hundreds of small stems.

Fosamine is an effective herbicide for small saplings and root suckers when applied to the foliage between July and September. Foliage should be well covered for best results. Basal bark applications or basal injections of triclopyr can also be used to control this species. Although not completely effective, cut-stump treatment of young suckers or clones with 25% a.i. glyphosate can be helpful. Be sure to treat every tree in the clone.

SHRUBS

Dogwoods

Gray or panicled dogwood: *Cornus racemosa*
Redosier or red twig dogwood: *C. stolonifera,* syn. *C. sericea*

There are other species of native dogwoods, but these are the two that are sometimes overly aggressive. Gray dogwood produces massive clones in prairies, while redosier dogwood is most aggressive in wetlands. These deciduous members of the Dogwood family provide excellent wildlife cover and nectar; fruits, leaves, and twigs are important sources of food for birds and other wildlife. Berries provide an important energy boost for migrating birds in the fall. Both species are used in landscaping and to stabilize streambanks.

A group of gray dogwood with their distinctive white berries on red stems.

Redosier dogwood is used for landscaping because of its bright red branches.

Habitat: Thickets, roadsides; prefer partial sun but will also grow in full sun.
  Gray dogwood: Prairies; grows on a variety of soils.
  Redosier dogwood: Wet to well-drained soils; prefers moist soils of damp woodlands, woodland and stream edges, swamps, marshes, and flood plains.
Height and Form: Typically 6–12′.
  Gray dogwood: Elliptical to egg-shaped; width typically two-thirds of height.
  Redosier dogwood: Globular or mound-shaped; occasionally to 20′ wide.
Leaves: Simple; opposite; elliptic; without lobes or teeth; usually with 3–5 pairs of veins.
  Gray dogwood: Blue-green, pale beneath, turning maroon purple in fall; 1–5″ long.
  Redosier dogwood: Deep green, pale beneath, turning orange or red and darkening to purple in fall; red leaf stalks; 2–4″ long; widely spaced veins.
Flowers: White; small; four petals.

Young leaves of gray dogwood.

Gray dogwood in bloom; June–July.

Redosier dogwood blooms in May and June.

Gray dogwood: In upright, cone-shaped spikes with red flower stalks; bloom June to July.

Redosier dogwood: In flat-topped, branched clusters that are 1–2″ wide with 8–12 flowers each; bloom in May and June.

Fruit: Small; white; round; berrylike drupes.

Gray dogwood: In cone-shaped clusters with red stalks; ripen in August.

Redosier dogwood: In flat clusters; ripen in mid-summer.

Stems: Multiple; form clumps.

Gray dogwood: Gray or brownish; interior is light brown.

Redosier dogwood: Bright red; interior is white when young.

Reproduction: By seed and root expansion.

Control: Controlled burns every few years provide good management for these species. Glyphosate, which is nonselective, can be carefully applied with wicks to foliage in August. (Use glyphosate approved for use in wet areas, if appropriate. Permits may be necessary.) Cut-stump treatment in late summer and fall with glyphosate or basal bark applications with triclopyr formulated for use with penetrating oil are also effective on gray dogwood.

Sumacs

Smooth sumac: *Rhus glabra*
Staghorn sumac: *R. typhina*, syn. *R. hirta*

Except for their differences in height, these two members of the Sumac family are very similar in appearance. They sprout easily and grow rapidly, forming circular thickets. They can dominate prairies that have not been burned in several years, shading out other species.

The fruits are an important winter survival food for deer, rabbits, and many kinds of birds. Sumac clones are helpful for naturalizing steep slopes

Female sumac colonies produce dark red, showy fruit spikes.

but are not recommended for planting in small urban gardens. Some people, especially those sensitive to poison sumac (*Toxicodendron vernix*) or poison ivy (*T. radicans*), may react to the juice of sumacs. Wear gloves and long sleeves when working with these species. Staghorn sumac gets its common name came from its hairy twigs that resemble deer antlers "in velvet."

Smooth sumac flower cluster and pinnately compound leaves.

> *Habitat:* Forest edges, prairies, clearings, disturbed sites, old fields, fencerows, roadsides; requires full sun; prefers sandy soil.
>
> *Height and Form:* Irregular form; narrow at the base, broad at the crown; spreading; rather picturesque, open crowns; thicket forming.
>
>> *Smooth sumac:* Usually 4–10′ tall; typically wider than it is tall.
>>
>> *Staghorn sumac:* 15–25′ tall, sometimes taller; as wide as it is tall.
>
> *Leaves:* Pinnately compound; alternate; 12″ long; 7–31 leaflets; leaflets are lance-shaped, saw-toothed, 2–5″ long, shiny green on upper surfaces and nearly white on lower surfaces; turn bright red in fall.
>
> *Flowers:* Yellowish green to white; less than 0.13″ wide; numerous; in much-branched, tight upright, 8″-long clusters; bloom late May and June; male and female flowers on separate trees or clones.
>
> *Fruit:* Showy, tight upright spikes; dark red; berrylike drupes; clustered in 6–10″ spikes; stand out in winter landscape; develop only on female clones; each fleshy on the outside with a hard center containing a single seed; mature by September; persist into winter.
>
>> *Smooth sumac:* Have short hairs.
>>
>> *Staghorn sumac:* Have long hairs.
>
> *Stems:* Relatively short-lived; roots persist and form new stems that can grow 3′ in one year.
>
> *Twigs:*
>
>> *Smooth sumac:* Smooth; covered with a whitish film that can be wiped off.
>>
>> *Staghorn sumac:* Velvety.
>
> *Bark:* Light brown and smooth on young plants.
>
> *Reproduction:* By spreading root stocks and seeds.
>
> *Control:* Sumac should be controlled where it spreads aggressively, shading out desirable vegetation. It should not be eliminated in plant communities where it occurred in presettlement times or in ravines within prairie communities. To control sumac in native plant communities, stems should be cut in July or shortly after flowering. Resprouts should then be cut in August. This double-cutting may need to be repeated for several consecutive years to achieve adequate control of dense populations. Spot-treating cut stumps with 10 to 20% a.i. glyphosate will minimize resprouting. Burning stands in August will also help control mature plants. Any resprouts that follow will need to be cut. Early spring fires may actually increase sprouting and spread. Mowing every year with a sickle-bar in mid-to-late July can decrease density. For badly degraded sites, triclopyr, glyphosate (1 to 2% a.i.), or 2,4-D can be applied to green foliage according to label directions. Avoid contact with nontarget species.

Twigs of staghorn sumac are velvety, while twigs of smooth sumac are nearly hairless.

Willows
Salix spp.

(See chapter 4 for information on these species.)

Currant/Gooseberries
Ribes spp.

There are many species of *Ribes* in the upper Midwest, both native and naturalized exotics. Some are rare, others common only in specific habitats, while those that tend to be invasive are more widespread and adaptable to a variety of wooded habitats. Both currants and gooseberries act as intermediate hosts for transmitting white pine blister rust, a disease that can cause significant damage to white pines (*Pinus strobus*). They have been eradicated from some areas as a result. However, colonies of these deciduous shrubs do provide cover for wildlife. Birds and mammals eat the berries; deer, moose, and porcupines eat the twigs and bark. Berries are often used in jams and jellies.

Currants and gooseberries have maplelike leaves.

Habitat: Most commonly found in disturbed woodlands.
Height and Form: 1–5′; rounded circular form; spreading or straggling.
Leaves: Simple; alternate; maplelike or resembling a goose's foot; 3–5 lobes.
Thorns:
　Currants: Lack thorns, except for bristly black currant (*R. lacustre*).
　Gooseberries: Most have 1–3 slender thorns or bristles at the bases of leafstalks.
Flowers: Small; inconspicuous; along stems; male and female parts are in the same flower.
　Currants: White to yellow; numerous (5 or more), in elongated clusters.
　Gooseberries: Greenish to white; 1–5 in short clusters.
Fruit: Dark purple to black, some red or yellow/green; roundish; often bristly (eastern prickly gooseberry, *R. cynosbati*).
　Currant: Numerous (5 or more) in elongated clusters.
　Gooseberries: 1–5 in short clusters; green, changing to purple as they ripen.
Reproduction: By seed.
Control: Pull young plants manually, with a leverage tool, chains or ropes. Regular fire can be effective in reducing vigor. Cut stems twice a year for several years, or cut stems in fall and stump treat with glyphosate or triclopyr. Use triclopyr as a foliar spray after leaf-out in spring.

Raspberries/Blackberries
Rubus spp.

Since European settlement, grazing and the absence of fire have caused populations of raspberries and blackberries to increase considerably. These deciduous shrubs, which are members of the Rose family, can be quite aggressive, taking up a great deal of space. Their thorns can also make movement through an area almost impossible.

The many varieties of raspberries and blackberries make excellent summertime food for more than 150 species of birds and mammals. Their dense, prickly thickets or brambles also provide refuge for small mammals and birds, and can be important for erosion control.

Habitat: Woodland edges, fencerows, clearings.
  Blackberries: Require full sun; dry soil.
  Raspberries: Full sun or partial shade.
Height and Form: 2–6′; usually in vaselike clumps with arching stems.
Leaves: Palmately compound; alternate; usually having 3 or 5 leaflets with numerous teeth; leafstalks are prickly.
  Blackberries: Green beneath.
  Raspberries: Usually white beneath.
Stems: Erect, arched, or trailing over the ground; usually prickly or bristly; twigs are green or red; new stems (canes) are produced each year; canes usually die after flowering and fruiting in the second year (roots are perennial).
  Blackberries: Angular; upright or arching; not white-powdered.
  Raspberries: Round; usually arched and white-powdered; sometimes rooting at tips.
Flowers: White; bloom late spring and early summer; male and female parts on the same flower.
  Blackberries: 1″ wide; petals are longer than sepals; borne in elongated clusters.
  Raspberries: 0.5″ wide; petals are no longer than sepals.
Fruit: Juicy berries; in tight clusters.
  Blackberries: Black.
  Raspberries: Black or red; white projections are left on stalks when berries are removed; a distinctive hole results at the base of fruits.
Reproduction: By seed and sometimes by stems rooting when in contact with the ground.
Control: Site can be burned or mowed several years in a row. Skipping a year may increase their spread. Mowing several times a year will reduce vigor. In small areas, the site can be mowed followed by a foliar herbicide treatment of resprouts using triclopyr, metsulfuron-methyl (both are broadleaf specific) or nonselective glyphosate. Raspberries will decline in forested sites as the forest develops and shade increases.

Raspberry and blackberry flowers are large, white, and occur in clusters.

Northern Prickly Ash
Zanthoxylum americanum

Northern prickly ash is a member of the Citrus family. A heavily barbed, upright shrub, it forms loose clonal thickets with shallow roots, especially in woodlots and old fields that have been grazed. Historically, prickly ash has had numerous medicinal uses including chewing leaves, bark, or fruits as a treatment for toothaches. Provides winter cover for wildlife, although they do not eat the berries. This plant is rare in parts of the southern and eastern United States.

Habitat: Rocky woods, old fields, riverbanks.
Height and Form: 4–10′ (sometimes 25′).
Leaves: Pinnately compound; alternate; 3–10″ long; have 2–6 paired leaflets with

Prickly ash has pinnately compound leaves that emit a lemony odor when crushed.

Global population growth and economic expansion contribute to ever-greater demands on natural ecosystems, on agriculture, and on governmental institutions. Greater U.S. demand for particular kinds of foreign imports generates new and more heavily used pathways for accidental introductions.... Clearing land often eliminates indigenous vegetation and creates pathways for invaders; more recreational visitors to natural areas increases the likelihood that harmful [non-indigenous species] will invade them. (U.S. Congress, Office of Technology Assessment, *Harmful Non-Indigenous Species in the United States,* September 1993)

an additional leaflet on the end; leaflets are finely toothed, egg-shaped to oblong, and hairy when young; leafstalks are often prickly; buds are rust colored and hairy; leaf scars have three dots within; leaves have a lemony odor when crushed.

Thorns: Large; paired; flank leaf scars and buds.

Flowers: Yellowish green; very small; in clusters without stalks; appear in spring before leaves are fully developed; male and female flowers are on separate plants.

Fruit: Pods; red and speckled, turning brown with age; small; clustered; dry; contain 1–2 shiny black seeds; scented.

Stems and branches: Prickly.

Bark: Citrus scent.

Reproduction: By seed and spreading rhizomes.

Control: Can be cut several times a year to reduce vigor. Stems are vulnerable to fire, so regular fire will provide good control. If fuels are insufficient or a prescribed, a propane torch can be used to spot burn stems. Be sure to treat all stems in a clone. Populations usually decline as a forest matures.

Vines and Vinelike Herbaceous Plants

Hog Peanut

Amphicarpaea bracteata, syn. *Glycine bracteata*

As its name implies, hog peanut is a member of the Legume family. A perennial, it is quite common in woods where its sprawls along the ground or twines tightly around other plants. Aboveground flowers produce pea-type pods, while small flowers at the base of the stems produce underground fruits. Birds eat both kinds of fruit; mice are known to cache its seeds. It can be a robust nuisance in small restorations or woodland and prairie gardens.

Habitat: Moist to dry thickets or woods, also wet meadows, savannas, and prairies. Especially common where sites are disturbed and somewhat sunny, such as trails and forest edges.

Leaves: Compound; alternate; divided into 3 egg-shaped, pointed leaflets; leaflets are 0.75–3″ long; light green and lack tendrils.

Flowers: Two kinds: (1) Pale lilac, pale purple, or white: narrow; pealike; 0.5″ long; two-lipped; hang in short, drooping clusters from leaf axils; (2) Small, inconspicuous flowers without petals near the base of the plant; bloom summer and early fall;

Fruit: From upper flowers flat, oblong to linear pods 0.67–1.5″ long and contain 3–4 mottled beans. From lower flowers fleshy, oval to pear-shaped, 1-seeded pods; sometimes located below the soil surface.

Stems: Twining; to 5′ long.

Reproduction: By seed.

Control: Pull repeatedly, several times a year for several years. Foliar application of clopyralid may be effective but must be done carefully to avoid damaging other legumes or composites.

Dodder

Cuscuta gronovii

Dodder, an annual in the Morning Glory family, is a parasitic plant that forms dense masses. Roots die soon after germination as tiny suckers become established on a host plant that supplies the dodder all its nourishment. This species, and several other dodders, are also destructive to some agricultural crops and can act as vectors for plant diseases. The other less common dodder species found in this region are difficult to distinguish from one another.

Habitat: Moist, low ground, and thickets.

Leaves: Absent or reduced to tiny scales.

Stems: Yellow or orange (being parasitic, the plant lacks green color); smooth; many-branched; twine tightly counterclockwise, like long, tangled spaghetti around and over other plants.

Flowers: White; tiny; 0.125″ long; bell-shaped; clustered; waxy; with 5 flaring lobes; bloom July to October.

Fruit: Small, papery seedpods, 0.125″ in diameter; contain 4–8 seeds.

Reproduction: By seed, which can remain dormant up to five years.

Control: The means to long-term control is to limit seed production. Control is difficult as the dodder is entwined with other plants and does not depend on regular roots to survive. Repeated pulling prior to flowering or fruiting will reduce seed production. It may be necessary to pull or clip the host plant and remove it from the site along with the dodder that is parasitizing it. A foliar application of 2,4-D will kill both the dodder and host plant.

Virginia Creeper or Woodbine

Parthenocissus quinquefolia

Virginia creeper is a perennial vine in the Grape family. Native to the eastern United States and southeastern Canada, it can be a groundcover or ascend 50′ into trees. Virginia creeper grows 3–6′ annually and is often planted as an ornamental. It is common in many deciduous forests of the Midwest and can become dominant in some sites. It provides fruit and cover for songbirds and other wildlife.

Characteristic 5-leaved whorls; sprawls across the ground and climbs trees; extensive, but shallow, root system.

Habitat: Full sun to partial shade, woodland edges, oak-hickory and bottomland forests, thickets.

Leaves: Palmately compound; alternate; normally 3–8″ across with long stems; divided into 5 coarsely toothed leaflets; leaflets are elliptical with toothed margins; can be 3″ wide and 6″ long; upper side of leaflets are green in summer, lower leaf pale; turn red in fall.

Flowers: Whitish or greenish; inconspicuous; in long-stemmed, branched clusters; bloom late spring and early summer.

Fruit: Bluish black berries; 0.25″ wide; on long-stemmed clusters; mature August to October.

Stems: Reclining along the ground or climbing tree trunks; slender; tendrils are found opposite the buds and can end in sticky pads, which help the vine cling to objects.

Reproduction: By rooting along the stems; by seeds.
Control: Fire can be used to top-kill stems and seedlings. Young plants can be pulled up and away from other plants. By following the vines to where they root, they can be cut and stump-treated with glyphosate or triclopyr in the late summer or fall. Foliar application of imazapyr can be effective but should be done carefully to avoid injuring nontarget plants.

Poison Ivy

Eastern Poison Ivy: *Toxicodendron radicans,* syn. *T. radicans* ssp. *negundo,* syn. *Rhus radicans*
Western or Northern Poison Ivy: *Toxicodendron rydbergii,* syn. *T. radicans* ssp. *rydbergii,* syn. *Rhus radicans* var. *rydbergii*

Poison ivy is a perennial plant that has evolved as a complex of two similar species and several subspecies. Eastern poison ivy is primarily found in the western, southern, and central parts of the Midwest. This is the form that can climb trees and is typically found in wet to mesic deciduous forests. Western or northern poison ivy is more widespread throughout the Midwest, and is found in rocky or sandy soils, near forest edges and trails, and on beaches and dry prairies. This species is usually a low-sprawling shrub, a trailing vine, or a small, upright shrub.

Although poison ivy can be overly abundant and dominating in some areas, it is generally not considered a serious threat to native plant communities. A member of the Sumac family, it provides valuable winter food for wildlife. Fifty species of birds are known to eat the berries without being harmed. Despite these ecological benefits, the oils from its leaves and stems can cause severe skin irritation and itching to about 85 percent of the human population. For this reason, poison ivy may need to be controlled in high-traffic areas.

A person's degree of sensitivity can vary over time. In the fall, dry poison ivy leaves continue to be poisonous. Smoke from burning poison ivy is also toxic to the skin and dangerous if inhaled. The presence of this species should therefore be a concern during prairie, savanna, and woodland burns.

To prevent exposure, learn to identify poison ivy and teach children how to identify it as well. Wear long pants, socks, and a long-sleeved shirt when walking in fields or wooded areas. Gloves are recommended when working with any plant in natural areas. Over-the-counter products are available to help protect the skin prior to exposure.

If you come in contact with poison ivy, wash the affected area with soap and water as soon as possible. Rubbing alcohol will also help remove the oil if applied soon enough. Dispose of gloves used to pull plants. Change clothes and wash them in hot soapy water separately from other clothing. Itching can be treated with cool compresses or over-the-counter preparations, such as calamine lotion. Contact your physician if the rash or itching is severe or involves the eyes, face or genitalia. (Exposure information supplied by the Waukesha Department of Health and Human Services, Waukesha, WI.)

Poison ivy in spring. Leaves may have reddish cast. Center leaflet extends from a long leaf-stalk.

Habitat: Open woods, woodland edges, river-bottom forests, dry prairies, beaches, parks, recreation areas, fencerows, roadsides, along trails.
Leaves: Compound; alternate; often glossy; with 3 strongly veined leaflets, each 2–4″ long; leaflets have pointed tips but are highly variable in appearance, even when borne on the same plant; elliptic to egg-shaped; may have toothed, lobed, wavy, or entire margins; middle leaflet extends from a long petiole, while lateral leaflets fasten directly to the main leaf stem or have very short stalks; droopy and reddish green in spring; become level and dark green when mature; turn yellow, orange, or bright red in fall.
Flowers: Off-white or greenish; small; 0.13″ wide; 5 petals; often hidden by the leaves; arise from leafless lateral branches in clusters of up to 25; borne in a 1–3′ long panicle in the leaf axil; dioecious; bloom May to July.

Poison ivy climbing a tree.

Berries: Yellowish white; round; hard; 0.13″ in diameter; develop on female plants in fall; often persist into winter allowing for identification without the leaves; eaten by birds and other wildlife.

Stems: Woody with gray bark; grow either horizontally along soil surface with upright leafy stalks (*T. rydbergii*) or as a climbing vine (*T. radicans* ssp. *negundo*); prominent leaf scars alternate on the vine and can help to identify poison ivy in winter months.

Roots: Multiple; sometimes aerial; attach to other plants and objects; cling to trees in summer and winter; hairy on older plants; allow for identification when dormant.

Reproduction: Primarily by seeds that are dispersed by birds and other wildlife;

Leaves may be dull or shiny.

Leaves turn yellow to red in fall. Margins may be wavy, toothed, or lobed.

Yellowish white berry clusters often remain into winter.

Aerial roots cling to trees.

also spreads by rhizomes and aboveground stems that often root where soil contact is made.

Control: It may not be necessary to control poison ivy unless it occurs near trails or other areas used by humans. When necessary, however, control efforts may need to be repeated for several years to deplete the seed bank. Uprooting individual plants is a common method of control. This is most safely done in late fall. Wear heavy gloves and protective clothing to prevent exposure. Poison ivy can also be controlled by repeatedly cutting, cultivating, or mowing plants. When pulling or digging poison ivy, the entire root should be removed and bagged to prevent resprouting. Poison ivy should never be burned or composted since human exposure to smoke or resprouting from composted material can occur. If using herbicides, a 1% a.i. solution of triclopyr can be applied to the foliage. Glyphosate or 2,4-D amine can also be applied to the foliage in late spring or early summer with a long-handled sponge or sprayer. These latter two herbicides may not be as effective as triclopyr and may require reapplication.

Wild Grapes

Summer Grape: *Vitis aestivalis* var. *aestivalis, V. aestivalis* var. *argentifolia*
Riverbank Grape: *V. riparia*
Fox Grape: *V. labrusca*

Grape vines can be very aggressive, climbing, covering, and shading out trees and other vegetation. They can be a major problem in lands set aside for timber production where they can twist and bend growing tree trunks and cause tree tops to break. The native summer grape (*V. aestivalis* var. *aestivalis*) is most common in southern areas of the upper Midwest, while northern summer

Grape vines often cover and shade out other vegetation.

grape (*V. aestivalis* var. *argentifolia*) is more common in the north. Fox grape (*V. labrusca*) appears to be native to the eastern part of the United States, but is introduced to the northwestern portion of the Midwest. Other grape species have the potential to be aggressive as well. Hybridization is common, resulting in multiple variations of these plants. Grapes are eaten in fall by many birds and mammals. Deer eat leaves and twigs. Some birds and squirrels also use the shredded bark to make nests.

> *Habitat:* Woodland edges, forest floors, open woods, fencerows, tree trunks; prefer some sun, usually growing toward it.
>
> *Leaves:* Simple; alternate; broad; shape variable, generally deeply heart-shaped at base; toothed; often 3-lobed; tendrils are often opposite leaves and buds.
>
> *Flowers:* Greenish; small; in branched clusters; bloom late spring to early summer.
>
> *Fruit:* Dark blue to black, globular berry; usually with powdery coating.
>
> *Stems:* Cling by forked tendrils; high climbing; dark colored with shreddy, peeling bark; have brown centers; lack thorns.

The dark fruit of wild grape often has a powdery cast.

> *Reproduction:* By seed and rooting where vines touch the soil. Vines resprout when cut.
>
> *Control:* Follow vines back to where they root and cut them at ground level, treating the cut stems with glyphosate or triclopyr formulated for dilution with penetrating oil as a basal bark treatment.

OTHER PLANT FORMS

Ragweeds

Common ragweed: *Ambrosia artemisiifolia*
Giant ragweed: *A. trifida*

Common ragweed and giant ragweed are annual forbs in the Composite family. They, and western ragweed (*A. psilostachya*) are found throughout the upper Midwest, especially on disturbed ground and in cultivated fields. Ragweed pollen, spread by the wind, is a major

Both giant and common ragweed (shown here) are a primary cause of hay fever.

White Snakeroot
Eupatorium rugosum

White snakeroot, a shade-loving, perennial herb in the Composite family, often forms dense stands in wooded settings. There are varying forms of this species due partly to hybridization. Cows that eat white snakeroot produce milk that may be fatal to humans if the milk is not processed or diluted with the milk of other cows. Fortunately, animals typically do not eat this plant unless other sources of food are scarce.

White snakeroot flowers, borne in branching, flat-topped clusters; bloom July to October.

Habitat: Rich woods, thickets, streambanks, disturbed soil.
Height: Typically 1–3′.
Leaves: Simple; opposite or sometimes whorled; somewhat heart-shaped; toothed; 2.5–7″ long; thin; lower leaves broadly egg-shaped; long-pointed; with slender, 0.75″ or longer stalks.
Flowers: Bright white; numerous; with small fuzzy heads on broad, branching, flat-topped clusters; individual flowers are about 0.2″ long and 0.16″ wide; bloom July to October.
Seeds: Black; 0.13″ long; with a tuft of white hairs.
Stems: Smooth; branched near top.
Roots: Many branched; fibrous.
Reproduction: By seed and short rhizomes.
Control: Plants easily pulled and generally do not resprout. Remove and destroy flower heads before seeds are produced.

Pale-Leaved Woodland Sunflower
Helianthus strumosus

Pale-leaved woodland sunflower is a perennial forb in the Composite family. It grows in dense patches and resembles several other native sunflowers, especially hairy sunflower (*H. hirsutus*). Birds and small mammals eat the seeds of this species and others in this genus, while muskrats eat the stems and foliage.

Habitat: Dry to wet woods, woodland edges, prairies, savannas, clearings, thickets, roadsides.
Height: Typically 2–7′.
Leaves: Simple; mainly opposite, although uppermost may be alternate; ovate to broadly lance-shaped; 3–8″ long; very rough; downy and pale beneath; finely toothed or without teeth; a pair of prominent lower veins run parallel to midrib; leaf stalks are winged; stalks of lower leaves are at least 0.25″ long.
Flowers: Have 9–15 yellow male ray flowers with yellow centers (female disc flowers); flower heads are 2.5–3.5″ wide; bloom August to September.
Fruit: Many; dark; hard shelled; flat at one end, pointed at the other; single ridge down middle of both sides; contain a single seed.
Stem: Smooth or slightly rough; often covered with whitish powder; branched near top.
Reproduction: By seed and creeping rhizomes.
Control: Pull or cut stalks several times a year until the population is depleted.

Jerusalem Artichoke
Helianthus tuberosus

Although its edible tubers are delicious in salads or roasted, Jerusalem artichoke is a problem in some areas because it forms dense patches. A very large, perennial forb and member of the Composite family, it shares many of the same characteristics as pale-leaved woodland sunflower, although it tends to grow on more disturbed sites. Its large leaves shade out other plants.

> *Habitat:* Disturbed ground, roadsides, fields, prairies, moist thickets, woods, riverbanks.
> *Height:* 6–10′.
> *Leaves:* Simple; lower leaves opposite; upper leaves alternate; large; 4–10″ long; egg- to lance-shaped; coarsely toothed; upper surface very rough and thick; lower surface hairy; have 3 main veins; leaf stalks are winged.
> *Flowers:* Have 10–20 yellow ray flowers with yellow disc flowers in center; flower heads are 2–3″ wide and have long, pointed bracts; bloom August to October.
> *Fruit:* Flat; wedge-shaped; often with black blotches; 1-seeded.
> *Stems:* Fuzzy or very rough; erect; usually branch near top.
> *Root system:* Comprised of fibrous roots from short rhizomes and immature, edible tubers.
> *Reproduction:* By rhizomes, tubers, and seeds.
> *Control:* Do not plant in a small prairie. Pull or cut stalks several times a year until the population is depleted. A selective herbicide can be applied prior to flowering.

Virginia Waterleaf
Hydrophyllum virginianum

Virginia waterleaf is a common, spring-blooming perennial herb in the Waterleaf family. Its young leaves can be eaten raw or cooked. It can be dominant in the understory of some woodlands, especially if they are disturbed.

> *Habitat:* Rich, wet to medium woods, thickets, floodplains.
> *Height:* Typically 6–24″.
> *Leaves:* Pinnately compound; mostly alternate; 2–5″ long; leaf stalks are divided with 5 or 7 egg- or lance-shaped, deeply toothed leaflets; usually appear water-stained with white blotches; smooth.
> *Flowers:* Bluish purple, pink or white; small, 0.33″ long;

Virginia waterleaf flowers are lavender or white; leaves have white blotches.

bell-shaped; closely clustered on tops of long stalks that rise above leaves; clusters nod; flowers have both male and female parts; stamens are long and protruding; bloom May to August.
> *Fruit:* Capsules containing 1–3 seeds.
> *Stems:* Somewhat smooth and weak.
> *Roots:* Fleshy; fibrous; from a very short to well-developed rhizome.
> *Reproduction:* By creeping rhizomes and seed.
> *Control:* Repeated pulling of plants. Spot treatment with a propane torch or with glyphosate can be effective.

Wood Nettle
Laportea canadensis

Wood nettle, a herbaceous perennial member of the Nettle family, is the only nettle with alternate leaves. It is found in every state from the East Coast to the Rocky Mountains.

> *Habitat:* Rich wet to medium woods and streambanks.
> *Height:* 16–40″.

Stems of wood nettle are bristly with skin-irritating, stinging hairs.

Leaves: Simple; alternate; egg-shaped; coarsely toothed; 3–6″ long; thin; leaf stalks are long.

Flowers: Greenish; very small; 0.16″ long; lack petals; grow on top and from leaf axils in loose, branching clusters; bloom July and August.

Fruit: Tiny, crescent-shaped seeds. ·

Stems: Stout; bristly with stinging hairs.

Reproduction: By seed.

Control: Cut with the use of a mower, scythe, or handheld weed whip. Plants can be easily pulled. If herbicide use is necessary on this species, 2,4-D amine is generally effective. Do not handle this plant without gloves and other appropriate clothing (including long pants if mowing or cutting with a weed whip); it contains an acid that causes severe, burning skin irritation.

Bracken Fern or Western Bracken Fern
Pteridum aquilinum

Bracken fern, a perennial member of the Bracken family, is one of the earliest ferns to appear in spring or after a fire; it provides quick vegetative cover after soil disturbance. It provides food and cover for a wide range of animals. One of the first plants top-killed by frost in fall, which results in a crisp, brown foliage, this fern is extremely common and often forms large colonies of nearly solid stands. It can be very competitive in disturbed or low nutrient soils. Allelopathic chemicals leached from dead ferns limit the germination and growth of some plants.

All parts of bracken fern may be toxic if consumed by livestock or humans for any extended period of time. In Japan, young leaves have been used as a food source and may be linked to increased incidence of stomach and other forms of cancer.

Habitat: Full sun to partial shade, woodlands, open pastures, hillsides, roadsides, thickets; prefers acidic soils.

Height: Typically 1.5–4′ (sometimes 7′).

Leaves: Curled and covered with silvery gray hair when emerging in spring; triangular when mature; typically reach 3′ long and 3′ wide; dark green; leathery; become almost horizontal; many branched with numerous oblong leaflets that are divided into narrow, blunt-tipped

Bracken fern typically grows 3′ tall, has triangular set of leaves, and spreads directly from rhizomes.

subleaflets; a continuous line of brownish spore cases are found on the underside edge of leaflets; cases are difficult to see (or are completely covered) because of inrolled leaf margins.

Leaf stalks: Rigid; tall (about the same length as the leaf); smooth; grooved in front; green to dark brown during the growing season; rise directly from the rhizomes.

Spores: Minute; brown; produced August through September.

Root system: Comprised of black, scaly 0.5″-thick rhizomes that can grow 20′ long and 10′ deep; roots are stout, black, wide-spreading, and grow sparsely along the rhizomes.

Reproduction: By spores and creeping rhizomes.

Control: Deplete food reserves in rhizomes by continually removing aboveground foliage after the fronds have fully expanded at least twice in the same growing season. If herbicide use is necessary, cut first, and apply glyphosate to the resprouts after they unfurl. English researchers are investigating biocontrol methods.

Cut-Leaved or Green-Headed Coneflower
Rudbeckia laciniata

Found in much of the United States and southern Canada, cut-leaved coneflower tends to form dense patches from rhizomes. A perennial member of the Composite family, it also provides seeds for birds and nectar for butterflies.

Habitat: Rich, low ground, pastures, wet prairies, sedge meadows, swamps, moist thickets and woods, streambanks.

Height: Typically 3–7′.

Leaves: Compound lower leaves; alternate; lower leaves with 3–7 irregular, deeply cut lobes; leaf stalks are long.

Flowers: Daisylike; 6–10 drooping yellow ray flowers with greenish yellow flowers in central cone-shaped disc; flower heads are 2–4″ in diameter; bloom July to September.

Seeds: Gray with 4 teeth on one end.

Stems: Leafy; branched; smooth; often whitish.

Root system: Dense with fibrous roots extending below rhizomes.

Reproduction: By rhizomes and seeds.

Control: Deplete food reserves in rhizomes by continually removing aboveground foliage. If herbicide use is necessary, apply glyphosate with a cotton glove (over a rubber glove), by grabbing the bottom of the plant and wiping the herbicide onto the stem and leaves.

Flower head of cut-leaved coneflower should not be confused with black-eyed Susan (Rudbeckia hirta), *which has a dark brown flower disk, or yellow coneflower* (Ratibida pinnata), *which has a more pointed disk.* Photo by Tom Brock

Cupplant
Silphium perfoliatum

A perennial forb of the Composite family, cupplant is often found in remnant and restored prairies of the upper Midwest. Birds, butterflies, insects, and small mammals drink rainwater held in the cup formed by this plant's upper leaves. Hummingbirds, bees, and butterflies also enjoy its nectar, and birds eat its seeds in the fall. While cupplant has many ecological benefits, it may not be appropriate for planting in small prairies where it can dominate and eliminate diversity.

Many, small flowers on long, stout stems; leaves clasp stems to form cups.

Habitat: Wet to medium prairies, fencerows, roadsides, open moist woods, meadows, streambanks, full sun to partial shade.

Height: 10′.

Leaves: Simple; opposite; larger leaves are 6–12″ long and 4–8′ wide; egg-shaped; pairs of upper leaves wrap around stems at their bases to form a cup; rough on both sides with saw-toothed edges.

Flowers: Numerous, 20–30 yellow ray flowers surround yellow disc flowers; flower heads are 2–3″ in diameter; bloom late summer and early fall.

Fruit: Gray to black, oblong sunflower-like shells; 1-seeded.

Stem: Stout; square; smooth; branching slightly at the top.

Reproduction: By seed and rhizomes.

Control: Deplete food reserves in rhizomes by cutting the stems at the base or mowing.

Goldenrods

Canada goldenrod: *Solidago canadensis* var. *scabra, S. canadensis* var. *hargeri*
Tall goldenrod: *S. gigantea*

Like the many other species of goldenrod, Canada and tall goldenrod are perennial forbs in the Composite family. They are very similar in appearance and may hybridize. Canada goldenrod can create clones that are 30′ or more wide. Both species are considered invasive in Europe. Goldenrods are often unfairly blamed for hay fever. Hay fever is primarily caused by ragweeds, especially giant ragweed, which bloom around the same time of the year. Wildlife use of these two goldenrods is low in comparison to their abundance.

Habitat: Prairies, roadsides, pastures, old fields, perennial croplands, ditches, riverbanks, floodplains, open woods; do not tolerate extremely wet or dry soils or shade.

Height:
   Canada goldenrod: 1–5′.
   Tall goldenrod: 2–7′.

Leaves: Simple; alternate; lance-shaped; sharply toothed; taper at both ends; without leaf stalks; all nearly the same size; midrib and two parallel lateral veins are prominent.

Canada goldenrod: Narrow; up to 6″ long; usually smooth on top, hairy beneath, especially on veins.

Tall goldenrod: Usually smooth or with soft hairs; up to 6″ long.

Flowers: Yellow; numerous; small; tightly packed at tops of stems in showy, elongated pyramid-shaped clusters; plumes are 3–9″ tall; flower stalks are arranged somewhat horizontally from a central axis and curve backward.

Canada goldenrod: Individual flower heads are less than 0.125″ long; bloom July to October.

Tall goldenrod: Individual heads are about 0.18″ long; bloom August through October.

Stems: Leafy; usually unbranched; slender.

Canada goldenrod: Softly hairy on upper half; hairless on lower half.

Tall goldenrod: Smooth; pale green or purplish; usually covered with a whitish coating.

Reproduction: By prolific seed production that is wind-dispersed, and by extensive rhizomes.

Control: Small patches can be pulled when the soil is damp, or flower heads can be cut off in peak bloom followed by a strong application of glyphosate to the cut surface. Mowing in midsummer can decrease abundance.

Canada goldenrod flowers are borne in showy, plumelike clusters; leaves are lance-shaped and heavily toothed.

Stinging Nettle
Urtica dioica

Stinging nettle, as its name suggests, contains an acid that causes a severe, burning skin irritation. Do not handle this species without gloves and other appropriate clothing for protection. In spite of its toxic affects, stinging nettle has many medicinal, culinary, and other uses. It also provides food for butterflies. Introduced varieties may be present. There are two perennial varieties: *U. dioica* var. *procera*, which is a native and widespread in the upper Midwest, and *U. dioica* var. *dioica*, which is from Europe and is not as common in the Midwest.

Habitat: Floodplains, moist open forests, roadsides, pastures, barnyards; usually found in damp, rich, disturbed soils but will tolerate dry soils and partial shade.

Height: 2–5′.

Leaves: Simple; opposite; saw-tooth edges; dark green; 2–4″ long; 0.75–1.25″ wide; covered with coarse, stinging hairs; leaf stalks are long.

U. dioica var. *procera:* Oval to lance-shaped leaves that are seldom heart-shaped at the base, with shallow (0.125–0.25″) teeth, stinging hairs are sparse and found primarily on the underside.

U. dioica var. *dioica:* More broadly oval leaves; a heart-shaped base, deeper (0.25–0.5″) teeth; densely hair on both sides.

Flowers: Greenish white; tiny; without petals; in slender, feathery, interrupted clusters in leaf axils; male and female flowers on separate plants on the European species; bloom June to September.

Fruit: Pods containing 1 seed.

Stems: Erect; square; slender; hollow; covered with stinging hairs; slightly branched near top. The European species has weak sprawling stems.

Leaves and stalks of stinging nettle are covered with skin-irritating, stinging hairs.

Reproduction: By seed and rhizomes.

Control: Nettles do not compete with grasses or in dense shade. Cut with a mower, scythe, or handheld weed whip. Plants can be easily pulled. If herbicide is necessary on this species, 2,4-D amine is generally effective. Do not handle this plant without gloves and other appropriate clothing (including long pants if mowing or cutting with a weed whip) since it contains an acid that causes severe, burning skin irritation.

GRASSLIKE SPECIES

Common Cattail
Typha latifolia

The native common cattail is often dominant in marshes, but generally not invasive in other wetlands unless there is disturbance. (See chapter 3 under "Cattails" for information on this species and its' non-native relatives.)

Ecological Restoration, Education, and Community Action

If suburbia were landscaped with meadows, prairies, thickets, or forests, or combinations of these, then the water would sparkle, fish would be good to eat again, birds would sing, and human spirits would soar. (Lorrie Otto, Wild Ones—Natural Landscapers, Ltd.)

DAVE EGAN, STEVE GLASS, AND EVELYN HOWELL

ECOLOGICAL RESTORATION

Invasive species management—as opposed to simple weed control and removal efforts—is a comprehensive, coordinated, and strategic approach conducted within an ecological restoration framework. The goals of restoration plans vary according to the situation but often include the enhancement of native species diversity, habitat improvement, or restoration of ecosystem processes. A restoration plan also contains an implementation schedule, a monitoring protocol, and provisions for adaptive feedback to improve performance.

Land managers sometimes use the phrase "invasive species removal" as a shorthand for the process of ecological restoration, and they often judge successful invasive species removal (or the kill rate) as equal to restoration success. But stopping at this point and hoping for the best is rarely sufficient to meet the goals of a restoration project. In fact, invasive species control and removal is just a small part of invasive species management—prevention, reversal of ecosystem disruption, and sustainable management—which, in turn, is a small part of the larger restoration picture.

According to the Wisconsin State Herbarium, in 2002 there were 2,437 taxa of native plants in Wisconsin and 795 taxa of introduced plants. Of the introduced taxa, 235 are well established and often widespread, 59 are locally established at particular places in small or large numbers, and 22 are persistent and spreading (generally moving out slightly from cultivation). (Composition of the Wisconsin flora, Wisconsin State Herbarium www.botany.wisc.edu/wisflora/composition .asp)

RESTORING NATIVE VEGETATION

Restoring native vegetation to a site that has been cleared of invasives requires studying the site conditions (soil chemistry, soil structure, hydrology, disturbance patterns) and, if necessary, returning them to their previous state. Planting appropriate native plant communities (e.g., prairie, savanna, woodland, wetland) on the site follows. Remediating the degraded site conditions denies invasive plants the environment that allowed them to thrive, and planting the cleared area with native plant community species provides the competition to keep invasives out or at very low levels.

Native plants, especially those that grow from corms, bulbs, or rhizomes, sometimes return on their own after invasives are removed, but in most cases the native species no longer exist in the soil seedbank. Simply reseeding or replanting the area is not as simple as it sounds. Restoring or naturally landscaping an area involves considerable planning and work prior to and after the planting to make it successful. This is especially the case in areas where disturbances to ecological processes, disruptions in ecosystem structure, or changes in soil chemistry or structure have occurred.

Most native insects prefer to eat their native host plants and, in fact, lack the enzymes needed to detoxify and digest non-native plants. Because birds rely heavily on insects as a source of high quality protein to nourish their nestlings, a reduction in insect numbers can have an adverse effect on birds and may be one of the factors causing declines in many bird populations. (Dr. Doug Tallamy, Department of Entomology and Applied Ecology, University of Delaware, 2002)

The restoration process involves the following series of interrelated steps.

- Analyze the site so that you will be able to decide what needs to be done to successfully carry out the restoration. This step involves determining, evaluating, and mapping existing soils, hydrology, drainage patterns, vegetation, adjacent land use, topography, and sun/shade patterns. You should also determine the historic processes and events that have produced the current conditions. There are many techniques available for determining the history of the land; in particular, see Egan and Howell's *The Historical Ecology Handbook: A Restorationist's Guide to Reference Ecosystems.* Using all this information, draw a base map that contains the size and location of the site's various attributes.

- Visit and inspect existing native plant communities—they, together with information from documents, such as the General Land Office surveys and other historic sources, provide models for your work. Study how they look as well as their vegetation structure and composition throughout the growing season. Make sketches and take notes. These areas are typically state natural areas, land owned by The Nature Conservancy, and other set-aside properties. Ask your state department of natural resources or local naturalists and botanists for locations and directions to these sites. Always obtain permission from the property owner before entering.

- Determine the goals and objectives of the restoration according to the information gathered from the site analysis and inventory, from studies of remnant models and from an assessment of the people or organization involved in the project. These goals and objectives describe the desired outcomes of the restoration activities, and serve as standards against which you can measure success. The goals and objectives should identify the plant community or communities that will be restored and should address such issues as the desired species composition, species abundance, species distribution patterns, community structure, and ecological processes. They should also reflect the ability of the landowner to monitor and manage the restoration once it is planted.

- Research is an essential part of restoration. A research plan should be developed early on in the planning process. Sometimes, especially when there is considerable uncertainty about a given aspect of the work, researchers will establish and evaluate "test plots" before restoring the rest of the site. In other cases, they may also establish different treatments within the restored area to test their hypotheses about restoration procedures or the native plant communities or ecosystem processes. The research plan may involve careful documentation of the restoration process, including what the environmental conditions were like before, during, and after the restoration. This information can be useful for management purposes and/or to generate hypotheses for future research and restorations.

- Create a list of the native species you plan to use including their size, flower color, blooming time, ease and type of germination, where they grow along the moisture continuum, plant community or communities they are typically found in, whether they grow in sun or shade, and the soil type they prefer. This information, much of which is available in native plant nursery catalogs, will help you match the species to the site conditions.

- Determine whether any of the existing site conditions (soil chemistry, soil structure hydrology, topography) will need to be changed and the ways and costs of doing so. This is a very important step because invasive weeds often signal a more basic problem with the land. There are many ways that site conditions can be modified to improve the success of a restoration. For example, gypsum can be added to soil whose structure has deteriorated. Some people have added sawdust to soil to ameliorate high levels of nitrogen. Other people have plugged drainage ditches to return the preexisting hydrologic conditions. Some conditions may be beyond the scope or budget of the project, and must simply be accommodated. This is especially true of continuing impacts, such as storm water or invasive plant populations, that come from adjacent properties and over which you have little or no control.

- Locate plant materials, soil amendments, consultants, and contractors as needed for the project. Use local ecotypes of seeds and plants whenever possible. Contact your local nature center, Wild Ones chapter, native plant organization, or state natural resources department for names of reputable nurseries. Do not dig plants from the wild; seed may be collected from the wild but only sparingly and with permission of the landowner.

- Determine what needs to be done to prepare the site for planting. Are there any invasive or otherwise competitive species left on site? If so, you must identify them and determine the best way to remove or control them. Does the soil or other planting medium need preparation? If so, you will need to determine if you can do it by hand, with a small machine, or a tractor. If the site requires extensive modification to its microtopography, as in wooded wetlands, low prairies, and riparian areas, extra planning and special machinery may be necessary. As is often the case in prairie restorations, shallow plowing helps release weed seeds in the seedbank, removing them from competition before the native species are planted. In woodland restorations that involve tree planting, it may be necessary to plant a shelterwood or nurse crop of fast-growing trees to provide shade for tree or herbaceous species that will be planted later.
- Decide on and diagram the planting scheme. Determine the numbers and proportions of species that will be planted. These decisions are typically derived from information gathered about the reference model, the site conditions, and project budget. Species are often identified and assembled as mixes (seeds) or groupings (plants and seeds). Individual or specimen plants are also used and should be identified.
- On a copy of the base map of your site, draw in the locations of individual plants, seed mixes, and groupings. This is your planting diagram.
- Plant the restoration. Planting a restoration depends on a number of factors—time of the year; availability of plant material, labor, and equipment; and the type of plant community being restored. For example, prairies are typically planted in either late spring or during the fall. Snow-seeding (tossing seeds on top of snow cover) is also practiced. Some prairies are planted by hand; other, usually larger, sites are planted with seed drills or seed spreaders drawn behind tractors or ATVs. Trees, shrubs, and plant seedlings may be planted by hand or by machine. Projects with large plants and/or large numbers of plants typically require the use of machinery. Watering, especially during droughty periods, often helps seeds germinate or young seedlings survive. Not all the species shown in the planting diagram need to be planted at the same time. Species can be planted in phases depending on how they respond to competition within the plant community. This technique is known as interseeding. A short-lived cover crop, such as annual rye or oats, is often added to prairie and savanna seed mixes to provide competition against the weeds and to control erosion as native species are becoming established. In cases where a more immediate visual or competitive effect is desirable, seedlings are used. The effect is offset, however, by the considerable expense of two-year-old plants.
- Following the planting, the site should be monitored to see if the restoration plans are being met. If they are not, you need to analyze what is taking place and decide on measures to correct the situation.
- Restoration sites typically need some type of management to either control invasive plants or reestablish ecosystem disturbance processes, such as periodic burning or flooding.

This is the general process for restoring a native plant community. It is considerably more complicated than simple invasive plant removal. There are many excellent sources of information about how to restore specific plant communities. See Appendix C, "Resources for Natural Landscaping and Ecological Restoration" for more information.

In a 1987 study conducted in Tucson, Arizona, research demonstrated that the density, species richness, and overall diversity of territorial native bird species corresponded closely to the volume of native plant species, but did not correspond to the volume of exotic plant species. The study suggested that native bird populations may be better retained in areas of urban development by landscaping with native plants in such a way as to retain predevelopment distributions of vegetation volume. (Based on the article, "Effects of urbanization on breeding bird community structure in southwestern desert habitats," G. S. Mills, J. B. Dunning Jr., and J. M. Bates, *Condor* 91:416–428)

In pushing other species to extinction, humanity is busy sawing off the limb on which it is perched. (Paul Ehrlich, Stanford University, Center for Conservation Biology, Stanford, CA)

Future generations of Americans deserve to inherit healthy productive wildlands, not vast landscapes infested with spiny, poisonous weeds that are unfit for people or wildlife. We must be wise enough and committed enough—right now—to increase our cooperative weed efforts at all organizational levels and to fully implement enough cooperative weed management areas so that the history of weed spread does not repeat itself over and over again across these public lands we value so highly. (Jerry Asher, "The Spread of Invasive Weeds in Western Wildlands: A State of Biological Emergency," Bureau of Land Management, Portland, OR 1998)

ELIZABETH J. CZARAPATA

EDUCATION

Human actions are the primary means for invasive species' introductions. Because we all share a certain amount of responsibility for invasive plant control, several ways to help limit their spread are listed below.

1. Learn how to identify and properly control invasive plants found in your area.

2. Tell your neighbors about invasive plants and how to control them.

3. Make sure invasive plant education is part of your school's curriculum. Let teachers know about resources available on this topic. (See Appendix A, "Resources for Information about Invasive Plants.")

4. Offer to do presentations about invasive plants for school classes.

5. Encourage teachers to get students out of the classroom and into the nature centers and parks where hands-on land management education can forever change the way they look at nature.

6. Contact local conservation organizations or the Master Gardeners (through the Cooperative Extension Service) for their help in educating others about invasive plants. Ask if fliers and brochures on invasive plants are available for public literature displays or distribution.

7. If there is no local conservation organization in your area that is willing to promote invasive plant education and control, consider starting your own. Incorporating and obtaining nonprofit status may be helpful. Grants may be available to support various projects. Ways to promote your group's effort include:

   - group T-shirts or caps
   - buttons
   - placing appropriate yard signs near invasive plant work sites to let people know that this is an official, organized event
   - distributing informative fliers in the neighborhood near work sites or posting them in public places
   - using a large banner to promote your organization for parades and educational displays or at work sites.

8. Set up educational displays at garden centers, poll-ing sites, hardware stores, street fairs, festivals, libraries, or municipal buildings.

9. Take photos of volunteers actively controlling invasive plants. Send the photos along with a press release to local newspapers. It is helpful to include contact information for experts on invasive plants and related brochures from respected conservation organizations.

10. Write letters to the editor of the local newspaper about the importance of controlling invasive plants.

11. Provide your library with educational materials on invasive plants, and let the community know where they are available.

12. Establish demonstration plots with corresponding signs to educate the public about how removal of invasive plants will affect the appearance and ecological health of an area. "Before and after" photos can be helpful.

13. Weather-resistant educational signs about invasive plants that are threatening your area can be discreetly placed on park trails or near a nature center.

14. Invasive plant awareness can be promoted at crucial times of the year. For example, April can be declared "Honeysuckle Awareness Month." Announce the event in local newspapers and explain why it is important. Create a display at the public library, municipal hall, or election site using photos of Eurasian honeysuckle and a bouquet of budding branches. Supply handouts that discuss why nonnative honeysuckle should be controlled, how to identify it, and control techniques. Similar displays can be set up in May for garlic mustard, in July for purple loosestrife (creating a bouquet of purple loosestrife may be illegal in some states due to concern about spread), and in October or November for Eurasian buckthorn.

15. Before beginning to remove invasive plants, especially trees and shrubs, educate neighbors about what you will be doing and why before you start. Provide them with educational materials from a reputable organization to back up what you are saying. If you fail to do this, it may be difficult to change their attitudes toward your efforts if neighbors or others become angry.

16. Encourage various civic organizations to invite speakers who are familiar with invasive plants. Slide presentations are helpful.

17. Use local cable stations or Web sites to alert citizens about invasive plants, or to inform them about related educational displays and events.

ELIZABETH J. CZARAPATA
COMMUNITY ACTION

There are two arenas of community action for those interested in controlling the spread of invasive plants—control efforts and policy changes. Find the one that best fits your interests and personality, and get involved.

Control Efforts

1. Contact land managers at your local nature center, parks department, conservation organization, natural resources department, or The Nature Conservancy and ask how you or your organization can help in the war on weeds. Land managers often have a limited staff and welcome volunteer help to locate and eradicate invasive plant infestations before they grow out of control.

2. "Adopt a spot." In many areas, invasive plants have gotten so out of hand that it is easy for someone trying to control these plants to be overwhelmed and get discouraged. One way to help prevent this and ensure that certain areas are cleaned out on a regular basis, and that seedbanks are being depleted, is to informally adopt a spot in an appropriate public natural area, perhaps located near your home. Be sure that land managers or public officials approve of your actions and understand what you are doing and why. Keep them informed.

3. You may want to get your family, club, church, or place of employment involved with "adopt a spot." Get expert recommendations about which plants pose the greatest ecological threat to the area and should be targeted for control first, what control techniques to use, what time of year is best to do

control work on a particular species, and where your work would be most beneficial. High-quality natural areas should be addressed before poor-quality areas that may not have a native seedbank remaining.

4. With permission from municipal officials, involve other groups such as high school students, civic organizations, Scouts, or businesses in invasive plant control efforts in local parks. Students may be able to use this time for community service hours, sometimes required for graduation, or for extra credit for science classes. Remember to keep volunteer safety in mind. Encourage volunteers to wear protective clothing and eye wear, point out poison ivy locations before weeding begins, and save steep slopes for the more agile, experienced weeders, or, if necessary, avoid them altogether.

5. Volunteer organizers may wish to name certain natural areas after particularly dedicated, long-term volunteers as a way of showing appreciation for their work, e.g., Myra's Ravine.

6. Financially support organizations that promote healthy natural areas.

Policy changes

1. To gain the support of municipal officials, inventory local parks for invasive plants and record your findings for presentation to the park board or similar committees. Encourage them to train and use park employees or professional services for invasive plant control—especially for those tasks unsuitable for volunteers (e.g., use of chain saws to cut large trees or shrubs or broadcast spraying of herbicides).

2. Insist that government officials support adequate funding for invasive plant education and control.

3. Organize tours for government officials to show the impact of invasive plants firsthand. Taking them to a healthy natural area, then to one overrun with weeds, can reveal a dramatic difference in the area's quality, diversity, and beauty.

Appendix A

Resources for Information about
Invasive Plants

Prairies

So rare,
So taken for granted,
So splendid, diverse, and unique,
So vital to the survival of countless wild creatures,
So poorly recognized and understood,
So easily violated and destroyed.

"Prairies"—Elizabeth J. Czarapata

BOOKS/BOOKLETS/JOURNALS/MANUALS

Biological Invasions, an academic journal; www
.kluweronline.com/issn/1387-3547.

*Biological Pollution: The Control and Impact of Invasive
Exotic Species.* 1992. Bill N. McKnight, ed. Indiana
Academy of Science. Indianapolis, IN. IAS
Publications, 1102 North Butler Ave., Indianapolis, IN
46219. Features twenty-one presentations from a 1991
symposium on invasive species. $30 plus $2.50 for
shipping and handling.

Chicago Wilderness publications:
- *Biodiversity Recovery Plan* provides a consensus
  assessment of the current state of biodiversity,
  threats, and actions needed to solve problems.
- *Chicago Wilderness Biodiversity Atlas* shows the substantial biological riches and heritage that remain in
  this highly urbanized region.
- *Protecting Nature in Your Neighborhood,* by the
  Northeastern Illinois Planning Commission, pres-
  ents the many things that local government and
  individuals can do to implement the Biodiversity
  Recovery Plan and protect and restore biodiversity.

Common Weeds of the United States. 1971. Agricultural
Research Service of the United States Department of
Agriculture, Dover Publications, NY.

Harmful Nonindigenous Species in the United States. 1993.
U.S. Congress, Office of Technology Assessment,
Washington, DC 20402-9328. Summarizes the
negative impacts of introduced plant and animal
species in the U.S. and offers suggestions on
addressing the issue. OTA Publications Office, U.S.
Congress, Washington DC 20510-8025. (202) 224-
8996. $21.

Illinois Vegetation Management Manual, Illinois Nature
Preserves Commission. (217) 785-8686, www
.inhs.uiuc.edu/edu/VMG/VMGintro.html.

*Invasive Plants, Changing the Landscape of America: Fact
Book.* Randy Westbrooks, 1998. Federal Interagency
Committee for the Management of Noxious and Exotic
Weeds (FICMNEW), Washington, DC 109 pp. Order
from U.S. Government Bookstore, Reuss Federal Plaza,
Suite 150W, 310 W. Wisconsin Ave., Milwaukee, WI
53203. (414) 297-1304, email: milwaukeebks@ibm.net.

Invasive Plants of the Southern Tier. Carol Estes Mortensen.
United States Department of Agriculture, Forest Ser-
vice Region 9, 2002 edition. Booklet. Contact Leech
Lake Division of Resources Management, 6530 Hwy. 2
NW, Cass Lake, MN 56633.

Invasive Plants: Weeds of the Global Garden. Brooklyn
Botanical Garden, 1000 Washington Ave., Brooklyn,
NY 11225. $7.95 plus $3.75 shipping and handling.

Life Out of Bounds: Bioinvasion in a Borderless World. Chris
Bright, World Watch Institute, 1776 Massachusetts Ave.
NW, Washington, DC 20036. (800) 555-2028 or (202)
452-1999, www.worldwatch.org.

Minnesota Invasive Non-native Terrestrial Plants. Minnesota
Department of Natural Resources, Trails and Water-
ways Division. Revised 2002. Booklet. Contact the
Minnesota DNR Information Center, 500 Lafayette
Road, St. Paul. MN 55155, (651) 296-6157 or (800)
646-6367.

Nature Out of Place—Biological Invasions in the Global Age.
2000. Jason Van Driesche and Roy Van Driesche. Island
Press. www.islandpress.org.

Ontario Weeds, (publication # 505) contains line drawings
of 315 species and 28 pages of color plates. Publications
Ontario, 50 Grosvenor St., Toronto, Ontario M7A 1N8,
Canada (416) 326-5300. $15 plus $1.80 for shipping
and handling.

Prairie Future publications, contact Prairie Future Seed
Company, LLC at P.O. Box 644, Menomonee Falls, WI
53052-0644, (262) 820-0221. Ask for current prices and
shipping rates for:
- *Weed Control in New Native Meadows and Gardens*
- *Photo Library of Weeds Common to Native Meadow &
  Garden Plantings with Guidelines for Weed Control*
- *Weed Index Coefficient of Risk*

Weeds of Nebraska and the Great Plains. 1994. Nebraska Department of Agriculture, Bureau of Plant Industry, P.O.
Box 94756, Lincoln, NE 68509. (402) 471-2394. $25
(includes shipping).

Weeds of the Northeast. 1997. Richard H. Uva, Joseph C.

Neal and Joseph M. DiTomaso. Cornell University Press. An identification reference for nearly 300 species.

Weeds of the Northern Lakes States. 2000. Carol Estes Mortensen. Steve Mortensen, photographer. Booklet. Leech Lake Division of Resources Management. Leech Lake Band of Ojibwe and Region 9 of the U.S. Forest Service.

Weeds of the West, Revised editions. 2000. Tom Whitson. Western Society of Weed Science.

Wetland Restoration Handbook for Wisconsin Landowners, Wisconsin Wetlands Association, 222 S. Hamilton St., Suite 1, Madison, WI 53703. $5.

Wisconsin Manual of Control Recommendations for Ecologically Invasive Plants, 1997, 1998, Randy Hoffman and Kelly Kearns, editors. Wisconsin Department of Natural Resources, Bureau of Endangered Resources, P.O. Box 7921, Madison, WI 53707-7921, (608) 267-7758.

CD-ROMs

Invasive and Exotic Species Compendium, 2002. Features 93 journal articles on invasive and exotic species and references to relevant websites. To order, contact the Natural Areas Association, P.O. Box 1504, Bend, OR 97709.

Invasive Plants of the Eastern United States: Identification and Control provides information about the identification, distribution, and control options for 97 invasive tree, shrub, vine, grass, fern, forb, and aquatic plant species. It also includes lists of invasive species from the Federal Noxious Weeds list, state regulatory agencies, pest plant councils, and other organizations. To order, contact Richard Reardon, USDA Forest Service, Morgantown, WV; (304) 285-2566, rreardon@fs.fed.us or visit www.invasive.org.

Purge Spurge: Leafy Spurge Database V. 4.0, features research reports, abstracts, bulletins, conference proceedings and other articles, photos, maps, and illustrations on leafy spurge and its management. To order, contact TEAM Leafy Spurge, USDA/ARS, 1500 North Central Ave., Sidney, MT 59270, (406) 433-2020, info@sidney.ars.usda.gov.

1000 Weeds of North America features an interactive key for identifying invasive species and weeds as well as images of 140 grass or grass-like species and 860 broadleaf weeds. It is available for $54.95 (includes shipping) from the Weed Science Society of America, (800) 627-0629.

CURRICULUMS

(See the Center for Invasive Plant Management Web site for most of the materials listed below and additional teaching resources: www.weedcenter.org/education/k12.html.)

"Invasive Weeds Curriculum," by Elizabeth J. Czarapata, grades 5–12, The Park People of Milwaukee County Inc., 750 N. Lincoln Memorial Dr., Suite 301, Milwaukee, WI 53202,(414) 273-7275. Includes a game, homework reading material for high school students, history of the invasive weed problem, crossword puzzle, word search, numerous activities. A fee will be required to cover the cost of printing, shipping, and handling.

"A Kid's Journey to Understanding Weeds," Intermountain Agriculture Foundation, A Kid's Journey to Understanding Weeds Project, 417 E. Fremont, Laramie, WY 82072-3143, FAX (307) 745-8733.

Montana Weed Project Teacher's Handbook, Missoula County Conservation District, 3550 Mullan Road, Ste. 106, Missoula, MT 59808-5125, (406) 829-3395. Lists classroom activities for grades 5–12.

"Purple Loosestrife Curriculum," Wisconsin Department of Natural Resources. Wisconsin Department of Natural Resources at (608) 221-6349 or (608) 266-7012.

"The Purple Loosestrife Project," lesson plans for K–12. Michigan State University, Room 334 Natural Resources, East Lansing, MI 48824, www.miseagrant.org/pp/.

FACT SHEETS/BROCHURES/POSTERS

Limited quantities of fact sheets, brochures, and posters may be available through your local county extension office, natural resources department, or conservation organization. For example, a garlic mustard brochure by Paul Hartman and Sharon Morrisey (Publication 200) and a buckthorn brochure by Laura Jull (Publication HT 2001) are available from University of Wisconsin-Extension County Offices. A purple loosestrife brochure is available from the Wisconsin Department of Natural Resources (see also www.dnr.state.wi.us/org/land/er/invasive/info/loose2.htm). The Nature Conservancy in Indiana has two brochures: *Invasive Plants in Indiana* and *Landscaping with Plants Native to Indiana.* They can be downloaded from www.inpaws.org/plants.html or obtained by contacting TNC at (317) 951-8818.

VIDEOS

Invasive Weeds—A serious threat to our environment, The Park People of Milwaukee County, Inc., 750 N. Lincoln Memorial Dr., Suite 301, Milwaukee, WI 53202, (414) 273-7275 for current price. Produced in conjunction with the Wisconsin Public Broadcasting program, *Outdoor Wisconsin.* Discusses invasive weeds in general with a focus on garlic mustard and common buckthorn.

Plants Out of Place, ITV Corporate Communications, sixty-minute television special, discusses invasive weeds in the U.S., (888) 380-6500.

Web Sites

American Lands Invasive Species Program: www.american-lands.org/forestweb/invasive.htm.

Aquatic Nuisance Species Task Force: www.anstaskforce .gov/.

Brooklyn Botanic Garden, Inc., *Invasive Plants: Weeds of the Global Garden* www.bbg.org/gar2/topics/sustainable/ 2000sp_invasive.html.

Brown University, Center for Environmental Studies: Report on Economic Impact of Invasives: www.brown .edu/Research/EnvStudies_Theses/full9900/mhall/ IPlants/Economic_Impact.html.

Bureau of Land Management, "Invasive Weeds: A Growing Pain" www.blm.gov/education/weed/weed.html.

California Exotic Pest Plant Council, "Invasive Plants of California's Wildlands," www.caleppc.org/.

Canadian Botanical Conservation Network: www.rbg.ca/ cbcn/en/projects/invasives/invade1.html and www .rbg.ca/newsletter/archives/0101/0101invasives.html.

Center for Aquatic and Invasive Plants: aquat1.ifas.ufl.edu/.

Center for Integrated Pest Management—Biocontrol Virtual Information Center: cipm.ncsu.edu/ent/ biocontrol/.

Center for Invasive Plant Management, K–12 teaching resources: www.weedcenter.org/.

Chemical and Pharmaceutical Press, Inc., *Greenbook Home,* a book with pesticide product information (labels, supplemental labels, and Material Safety Data Sheets) provided by pesticide companies: www.greenbook.net/.

Chicago Botanic Garden: www.chicagobotanic.org/ research/conservation/invasive.html.

Chicago Wilderness, Biodiversity Recovery Plan: www .chicagowilderness.org/biodiversity/threats/index.cfm.

Connecticut Invasive Plant Working Group: www .hort.uconn.edu/cipwg/.

Cornell University, Biological Control of Non-Indigenous Species: www.invasiveplants.net/.

Crop Data Management Systems, Inc., provides herbicide labels and Material Safety Data Sheets (MSDS): www .cdms.net/.

Environmental Protection Agency, Office of Pesticide Programs: www.epa.gov/pesticides/.

Extoxnet, a cooperative effort by five universities to provide objective, science-based information about pesticides; written for the non-expert: ace.orst.edu/info/extoxnet/.

Federal Interagency Committee for the Management of Noxious and Exotic Weeds: ficmnew.fws.gov/.

Florida Exotic Pest Plant Council: www.fleppc.org.

Forest Preserve District of DuPage County, Illinois, Habitat Improvement Project, restoring natural character to the landscape: www.dupageforest.com/ CONSERVATION/hip.html.

Great Lakes Indian Fish and Wildlife Commission, Exotic Plant Information Center, Upper Great Lakes Weed Mapping Project: www.glifwc.org/epicenter/.

Great Lakes Information Network: www.great-lakes.net/ envt/flora-fauna/invasive/invasive.html.

Herbicide-resistant weeds and their impact, international survey: www.weedscience.org/in.asp.

Illinois Native Plant Society: www.ill-inps.org/.

Illinois Natural History Survey: www.inhs.uiuc.edu.

Illinois Nature Preserves Commission, "Vegetation Management Manual": www.inhs.uiuc.edu/edu/VMG/ VMGintro.html.

Indiana DNR, Invasive Species: www.in.gov/dnr/ invasivespecies/.

Infrared Weeders: www.pesticide.org/RadiantHeatWeeders .pdf.

Invaders Database System—Invasives in Northwestern United States, Noxious Weed Lists for all states: invader.dbs.umt.edu/.

Invasive Exotic Pest Plants in Tennessee: www.se-eppc.org/ states/TN/TNIList.html.

Invasive Plant Association of Wisconsin: www.ipaw.org/.

Invasive Plant Atlas of New England: invasives.eeb.uconn .edu/ipane/.

Invasive Plant Council of New York State: www.ipcnys.org/.

Japanese Knotweed Alliance: www.cabi-bioscience.org/ html/japanese_knotweed_alliance.htm.

Kansas Department of Agriculture, Plant Protection and Weed Control Program: www.accesskansas.org/kda/ Plantpest/index.htm.

Kansas State University Research and Extension: www .oznet.ksu.edu/pesticides-IPM/pesticides/Pesticide websites pamphlet.pdf.

Kentucky's Exotic Pest Plant Council: www.se-eppc.org/ states/KY/KYlists.html.

Land Conservation Department, Monroe County, Wisconsin (factsheets): www.co.monroe.wi.us/.

"Methods of Introduction of Nonnative Plants into New Habitats: A Review," Jennifer Forman: www.massscb .org/epublications/fall2001/invasives.html.

Michigan Invasive Plant Council: forestry.msu.edu/mipc/.

Mid-Atlantic Exotic Pest Plant Council: www.ma-eppc.org/.

Midwest Invasive Plant Network: www.mipn.org.

Minnesota Department of Agriculture, Minnesota Noxious Weed Program: www.mda.state.mn.us/appd/weeds/ fsmnwp.html.

Minnesota Department of Natural Resources' Harmful Exotic Species Program: www.dnr.state.mn.us/ecological_services/exotics/index.html.

Missouri Botanic Gardens, "Linking Ecology & Horticulture to Prevent Plant Invasions": www.mobot.org/invasives/.

Missouri Flora Website: www.missouriplants.com/index.html.

Missouri Vegetation Management Manual: www.conservation.state.mo.us/nathis/exotic/vegman/.

National Association of Exotic Pest Plant Councils: www.se-eppc.org/nationaleppc.cfm.

National Biological Information Infrastructure: invasivespecies.nbii.gov/.

National Fish and Wildlife Foundation, fosters private, state, and local funding through challenge grants to conserve fish, wildlife, plants, and habitats on which they depend: www.nfwf.org/programs/grant_apply.htm.

National Invasive Species Council: www.invasives.org.

National Park Service, Alien Plant Invaders of Natural Areas fact sheets: www.nps.gov/plants/alien/factmain.htm.

National Park Service, *Handbook for Ranking Exotic Plants for Management and Control:* www.nature.nps.gov/publications/ranking/index.htm.

The Nature Conservancy, Weeds On The Web, photos, species identification, extensive summaries of selected species, control methods, tools: tncweeds.ucdavis.edu/.

NatureServe Explorer, an online encyclopedia, is the source for authoritative conservation information on more than 50,000 plants, animals, and ecological communities of the U.S. and Canada: www.natureserveexplorer.org/.

Nebraska Department of Agriculture, Nebraska Noxious Weed Program: www.agr.state.ne.us/division/bpi/nwp/nwp1.htm.

New England Invasive Plant Group: www.se-eppc.org/states/newengland.cfm.

New York Botanical Garden's Catalog of Invasive Plant Species in the Unites States: www.nybg.org/bsci/hcol/inva/.

North American Weed Management Association: www.nawma.org/.

Ohio Division of Natural Areas and Preserves, Invasive Plants page: www.ohiodnr.com/dnap/invasive/.

Ontario Vegetation Management Association: www.ovma.on.ca/.

Oregon Public Broadcasting, *American Field Guide* (includes videos): www.pbs.org/americanfieldguide/topics/plants/index.html.

Pacific Northwest Exotic Pest Plant Council: www.wnps.org/eppclet.html

The Park People, Friends of Milwaukee County Parks, "Weed-Out" volunteer weed removal program in natural areas of public parks: www.theparkpeople-milwaukee.org.

Pesticide Action Network Pesticide Database: www.pesticideinfo.org.

Plant Conservation Alliance, "Weeds Gone Wild": www.nps.gov/plants/alien/.

Prairie Enthusiasts, dedicated to preserving and restoring prairies of Wisconsin and northern Illinois: www.theprairieenthusiasts.org/.

South Carolina Aquatic Nuisance Species Program: www.dnr.state.sc.us/water/envaff/aquatic/.

Southeast Exotic Pest Plant Council: www.se-eppc.org.

Tree Trust. "Buckthorn Bust, Neighborhood Guide." City of St. Paul, Division of Parks and Recreation, and Minneapolis Park and Recreation Board. www.ci.stpaul.mn.us/depts/parks/environment/buckthorn/buckthorn_index.htm.

United Kingdom's Loughborough University, proceedings of international conferences about invasives: www.lboro.ac.uk/research/cens/invasives/index.htm.

United States Department of Agriculture, Invaders Database System: invader.dbs.umt.edu/Noxious_Weeds/.

United States Department of Agriculture, Vascular Plants Database: www.plants.usda.gov/.

United States Department of Transportation, *Roadside Use of Native Plants:* www.fhwa.dot.gov/environment/rdsduse/index.htm.

United States Environmental Protection Agency, "Weed laws and natural landscaping": www.epa.gov/grtlakes/greenacres/weedlaws/index.html.

United States Fish and Wildlife Service: contaminants.fws.gov/Issues/InvasiveSpecies.cfm.

United States Forest Service, biological control: www.fs.fed.us/foresthealth/technology/bcprog.shtml.

United States Geological Survey, Alien Plant Ranking System Version 5.1: www.npwrc.usgs.gov/resource/2000/aprs/aprs.htm.

United States Geological Survey, Nonindigenous Aquatic Species Distribution Database: nas.er.usgs.gov/plants/.

University of Florida Center for Aquatic and Invasive Plants: aquat1.ifas.ufl.edu/.

University of Florida Extension: edis.ifas.ufl.edu/TOPIC_Invasive_Weeds.

University of Georgia, Invasive Exotic Species: www.invasive.org/.

University of Minnesota, Minnesota Sea Grant: www.seagrant.umn.edu/exotics/fieldguide.html.

University of Wisconsin Extension: ipcm.wisc.edu/uw_weeds/extension/weedprofiles.htm.

University of Wisconsin–Madison Arboretum, Earth Partnership Program, teaching restoration in schools:

wiscinfo.doit.wisc.edu/arboretum/
earth_partnership_index.htm.
University of Wisconsin–Madison Herbarium, photos of
1,990 Wisconsin native and naturalized vascular plants,
includes distribution maps: www.botany.wisc.edu/
herbarium/.
Virginia Tech Weed Identification Guide: www.ppws.vt
.edu/weedindex.htm.

Weed Science Society of America, agricultural weeds and
links: www.wssa.net/.
Wisconsin Department of Natural Resources, invasive plant
manual, control methods, and links: www
.dnr.state.wi.us/org/land/er/invasive/index.htm.
The World Conservation Union (IUCN), Invasive Species
Specialist Group book, *100 of the World's Worst Invasive
Alien Species:* www.issg.org/booklet.pdf.

Resources to Help with General Plant Identification

The following publications are helpful tools when learning to identify plants in natural areas of the upper Midwest:

Borman, S., R. Korth, and J. Temte with illustrations by C. Watkins. 1998. *Through the Looking Glass: A Field Guide to Aquatic Plants.* Madison: University of Wisconsin Press.

Courtenay, B., and J.H. Zimmerman. 1978. *Wildflowers and Weeds.* New York: Simon and Schuster. (out of print)

Crow, G. E., and C. B. Hellquist. 2000. *Aquatic and Wetland Plants of Northeastern North America, Volumes I and II.* Madison: University of Wisconsin Press.

Dirr, M. A. 1997. *Dirr's Hardy Trees and Shrubs: An Illustrated Encyclopedia.* Portland, OR: Timber Press.

Gleason, H. A., and A. Cronquist. 1991. *Manual of Vascular Plants of Northeastern United States and Adjacent Canada.* New York: New York Botanic Garden.

Hightshoe, G. L. 1988. *Native Trees, Shrubs and Vines for Urban and Rural America.* New York: Van Nostrand.

Hightshoe, G. L., and H. D. Groe. 1998. *North American Plantfile.* New York: McGraw-Hill.

Holmgren, N. H., P. K. Holmgren, H. A. Gleason, and A. Croquist. 2004. *Illustrated Companion to Gleason and Cronquist's Manual.* Bronx, NY: Brooklyn Botanic Garden.

Ladd, D., and F. Oberle. 1995. *Tallgrass Prairie Wildflowers.* Helena and Bozeman, MT: Falcon Press Publishing, Inc.

Mohlenbrock, R. H. 2001. *Sedges: Cyperus to Scleria, Second Edition.* Carbondale, IL: Southern Illinois University Press.

Newcomb, L. 1977. *Newcomb's Wildflower Guide.* Boston: Little, Brown and Co.

Peterson, R. T., and M. McKenny. 1968. *A Field Guide to Wildflowers-Northeastern/North Central North America.* Boston: Houghton Mifflin Co.

Petrides, G. A. 1972. *A Field Guide to Trees and Shrubs.* Boston: Houghton Mifflin Co.

Petrides, G. A., and J. Wehr. 1988. *A Field Guide to Eastern Trees.* Boston: Houghton Mifflin Co.

———. 1998. *Eastern Trees.* Boston: Houghton Mifflin Co.

Royer, F., and R. Dickinson. *Weeds of the Northern U.S. and Canada.* Edmonton: University of Alberta Press.

Symonds, G. W. D. 1963. *Shrub Identification Book.* New York: William Morrow & Co.

———. 1973. *Tree Identification Book.* New York:William Morrow & Co

Thieret, J. W., and N. C. Olmstead. 2001. *National Audubon Society Field Guide to North American Wildflowers-Eastern Region.* New York: Random House.

Individuals who may be helpful for plant identification include horticulturists at Cooperative Extension offices, university botany departments or herbaria, and naturalists at nature centers.

Appendix C

Resources for Natural Landscaping and Ecological Restoration

NATIONAL AND REGIONAL RESOURCES

Lady Bird Johnson Wildflower Center: www.wildflower.org.

Land and Water magazine publishes an annual "Buyer's Guide" listing companies that sell native plants. See the Web site: www.landandwater.com. To order, call (515) 576-3191.

Society for Ecological Restoration International, www.ser .org, provides a forum for restoration-related activities both globally and regionally through their various chapters.

U.S. Environmental Protection Agency has a guide to landscaping with native wildflowers and grasses: www.epa .gov/greenacres.

University of Wisconsin-Madison Arboretum: www.wisc .edu/arboretum, and the Arboretum's journal, *Ecological Restoration* (www.ecologicalrestoration.info), are excellent sources for information about restoring native plant and animal communities.

Wild Ones—Natural Landscapers, an organization of native landscapers with chapters in many states: P.O. Box 1274, Appleton, WI 54912-1274, www.for-wild .org.

STATE AND PROVINCE RESOURCES

Illinois: Illinois Department of Natural Resources, Division of Natural Heritage, (217) 785-8774. Request their booklet, "Prairie Establishment and Landscaping." Another helpful resource is *Flora of Illinois* by Robert Mohlenbrock.

Indiana: Indiana Department of Natural Resources, Division of Nature Preserves, (317) 232-4052. Request the brochure, "Landscaping with Plants Native to Indiana" by the Indiana Native Plant and Wildflower Society, www.inpaws.org.

Iowa: Iowa Department of Natural Resources, Prairies and Preserves Division, (515) 281-3891, for restoration questions and a list of prairie seed dealers. Many dealers will also be able to answer restoration questions. Books that may be helpful include *Vascular Plants of Iowa* by Eilers and Roosa, and *Trees and Shrubs of Iowa* by Vanderlinden and Farrar.

Michigan: Michigan Association of Conservation Districts has a helpful Web site with links to national and state organizations that promote the use of native plants and the removal of invasives, www.macd.org/nativeplants/ nplinks.html.

Minnesota: Minnesota Department of Natural Resources, Division of Forestry, (651) 296-4491, for restoration information and plant sources. The University of Minnesota Extension Service, College of Agriculture, Food and Environmental Sciences at (800) 876-8636 also has helpful publications: "Plants in Prairie Communities," "Establishing and Maintaining a Prairie Garden," and "Minnesota's Forest Trees."

Ohio: Department of Natural Resources, Division of Natural Areas and Preserves, (614) 265-6453, for a list of prairie seed sources and a list of recommended native plants. Also, contact the Ohio State University Extension (614) 292-1607. Request their bulletin # 865, "The Native Plants of Ohio" or visit the Web site: www .ag.ohio-state.edu/~ohioline/b865/index.html.

Ontario: Wetland habitat restoration and information, see the Web site: www.on.ec.gc.ca/wildlife/wetlands/ glwcap-e.cfm. To obtain a manual about prairie and meadow (including wet meadow) restoration, see the Web site: www.tallgrassontario.org/planting_ the_seed.htm. For the Society of Ecological Restoration Ontario chapter, see Web site: www.serontario.org/ index.htm.

Wisconsin: For a list of native plant nurseries and restoration consultants, contact the Bureau of Endangered Resources, Wisconsin Department of Natural Resources, P.O. Box 7921, Madison, WI 53707, (608) 266-7012. A booklet listing native woody plants for wildlife, "So What Should I Plant?" is also available from the Wisconsin DNR, publication # WM-223-98. For a free listing of eighty-four deciduous trees, shrubs, and conifers that are native to Wisconsin, send a self-addressed, stamped envelope to "Native Trees and Shrubs," Wisconsin Natural Resources magazine, P.O. Box 7921, Madison, WI 53707. This publication includes plant requirements (sun exposure, moisture, pH preference range, soil type), wildlife value, landscape uses, and ecological communities where the plants are normally found.

Books and Catalogs

Admiraal, A. N., M. J. Morris, T. C. Brooks, J. W. Olson, and M. V. Miller. 1997. *Illinois Wetland Restoration and Creation Guide.* Champaign: Illinois Natural History Survey, special publication no. 19.

Anderson, R. C., J. S. Fralish, and J. M. Baskin, editors. 2000. *Savannas, Barrens, and Rock Outcrop Plant Communities in North America.* New York: Cambridge University Press.

Brumback, W. E. 1986. *Propagation of Wild Flowers.* Framingham, MA: New England Wildflower Society.

Cochrane, T. S., and H. H. Iltis. 2000. *Atlas of the Wisconsin Prairie and Savanna Flora.* Madison: Wisconsin Department of Natural Resources.

Conservation Commission of Missouri. 1999. *Public Prairies of Missouri.* Jefferson City: Missouri Department of Conservation.

Diekelmann, J., and R. Schuster. 2003. *Natural Landscaping.* Madison: University of Wisconsin Press.

Egan, D., and E. Howell. 2001. *The Historical Ecology Handbook: A Restorationist's Guide to Reference Ecosystems.* Washington, DC: Island Press.

Harper-Lore, B. 2000. *Roadside Use of Native Plants.* Washington, DC: Island Press.

Henderson, R. A. 1995. *Plant Species Composition of Wisconsin Prairies: An Aid to Selecting Species for Plantings and Restorations Based Upon University of Wisconsin–Madison Plant Ecology Laboratory Data.* Madison: Wisconsin Department of Natural Resources.

Hoen, A., and A. Larson. 1999. *Begin With a Seed: The Riveredge Guide to Growing Wisconsin Prairie Plants.* Newberg, WI: Riveredge Nature Center.

Kurtz, C. 2001. *A Practical Guide to Prairie Reconstruction.* Iowa City: University of Iowa Press.

Ladd, D., and F. Oberle. 1995. *Tallgrass Prairie Wildflowers.* Helena and Billings, MT: Falcon Publishing Co.

Marinelli, J., editor. 1994. *Going Native: Biodiversity in Our Own Backyards.* Brooklyn, NY: Brooklyn Botanic Garden.

Mills, S. 1995. *In Service of the Wild: Restoring and Reinhabiting Damaged Land.* Boston: Beacon Press.

Mirk, W. 1997. *An Introduction to the Tallgrass Prairie of the Upper Midwest: Its History, Ecology, Preservation and Reconstruction.* Boscobel, WI: The Prairie Enthusiasts.

Packard, S., and C. F. Mutel, editors. 1997. *The Tallgrass Restoration Handbook for Prairies, Savannas, and Woodlands.* Washington, DC: Island Press.

Prairie Moon Nursery catalog. www.prairiemoon.com.

Prairie Nursery catalog. www.prairienursery.com.

Sauer, L. 1998. *The Once and Future Forest: A Guide to Forest Restoration Strategies.* Washington, DC: Island Press.

Sayre, F, editor. 1999. *Recovering the Prairie.* Madison: University of Wisconsin Press.

Smith, J. R., and B. S. Smith. 1980. *The Prairie Garden: 70 Native Plants You Can Grow in Town or Country.* Madison: University of Wisconsin Press.

Stevens, W. K. 1995. *Miracle Under the Oaks: The Revival of Nature in America.* New York: Pocket Books.

Svedarsky, D., M. Kuchenreuther, G. Cuomo, P. Buesseler, H. Moechnig, and A. Singh. 2002. *A Landowner's Guide to Prairie Management in Minnesota.* Crookston, MN: Northwest Branch and Outreach Center, University of Minnesota.

Thompson, A. L., and C. S. Luthin. 2004. *Wetland Restoration Handbook for Wisconsin Landowners,* 2nd edition. Madison: Wisconsin Department of Natural Resources.

Thompson, J. 1992. *Prairies, Forests & Wetlands: The Restoration of Natural Landscape Communities in Iowa.* Iowa City: University of Iowa Press.

Glossary

Adjuvant	A chemical added to an herbicide to increase its effectiveness or safety.
Allelopathic	Producing chemicals that impede the growth of competing species.
Alternate (leaves, etc.)	Plant parts growing individually along a stem; not opposite to each other.
Annual	A plant that germinates, blooms, produces fruit, and dies in one growing season.
Anther	The enlarged part of the stamen that produces pollen.
Aquatic plant	Plants that live or grow in water; either submergent, emergent, or floating-leaf plants.
Auricles (grass)	Leaf appendages located where the blade and sheath join.
Axil	The position where the leaf joins the stem.
Axillary	Additional stems; other than the primary stem; located in or arising from an axil.
Ballast	Any heavy material, such as dirt, water, or rocks, used to help stabilize an object.
Basal leaves	Leaves radiating around the stem at ground level.
Biennial	A plant with a two-year growth cycle; germinating the first growing season and flowering, producing fruit, and dying the second growing season.
Biological control	Plant control using natural enemies, usually from the plant's native range, to control a species; e.g., insects or diseases.
Bolt	To send up a flowering stalk.
Brackish	Slightly salty or briny water.
Bract	A modified leaf (not always green) found at the base of a flower or flower cluster.
Broadleaf	A plant that usually has leaf veins in a netlike pattern and a taproot; leaves that are not needlelike.
Calcareous fen	Rare wetland community in North America watered by mineral-rich, alkaline groundwater or seeps.
Cambium layer	The thin, slippery tissue layer between the bark and wood of a tree or shrub where new growth occurs.
Cane	A long, jointed woody stem.
Catkin	A soft, downy or scaly, long and slender cluster of small flowers without petals that grows on birches, willows, and alders.
Caudex	The woody base of a perennial plant that sends up new herbaceous stems each year to replace those that died the year before.
Chemical control	Using herbicides to kill or setback an unwanted plant species.
Clone (plants)	A group of plants that have been produced from a single ancestor, as with root suckering.
Compound leaf	Comprised of two or more similar leaflets arranged on either side of a common midrib; may be pinnately or palmately compound.

Control (weed)	To destroy the aboveground portion of a noxious weed, preventing the development and spread of viable seeds or other propagules, or to control the underground spread of a plant by rhizomes.
Cover crop	A crop (often annual rye or flax) that will die after frost but is sown to grow quickly and protect the soil from erosion or to cover the soil to prevent weed seeds from germinating.
Cultivar	A plant variety produced by selective breeding.
Cultivate	Intentionally maintaining a particular plant.
Deciduous	Shedding the leaves annually; refers to woody plants.
Dicot-specific	Affecting only broadleaf plants.
Diluent	Water or another liquid that is mixed with an herbicide to weaken the solution.
Dioecious	Having male and female flowers of the same species on separate plants.
Duff	Decaying plant material (fallen leaves, stems, etc.) on the ground.
Ecospecies	A group of plants or animals that have been modified, over time, by their environment into a distinct species.
Ecosystem	An intricate community of organisms and its environment.
Elliptic	Resembling a flattened, symmetrical circle that is widest in the middle and tapering equally toward each end.
Emergence	Refers to plants breaking through the soil or water surface.
Entire (leaf)	Leaf without lobes or teeth.
Ephemeral	Having a very short growing season, as in spring wildflowers.
Eradicate	To completely eliminate a plant population from an area so that no further individuals of that species are present to emerge from seeds or rhizomes.
Exotic (plants)	Primarily European and Asian species that have been intentionally or accidentally introduced to North America. Also known as "alien," "non-native," or "non-indigenous" species. More technically, any plant existing outside its normal range.
Fen	A rare type of wetland that receives water from a continuous source of water (groundwater, springs, surface water) creating alkaline soil and supporting distinct vegetation.
Fern ally	A non-flowering plant that, like a fern, reproduces by spores, e.g., horsetails, club mosses, quillworts.
Fibrous (roots)	Threadlike and dense.
Fire-adapted plant community	A plant community, such as a prairie, that historically existed with fire and will thrive with occasional burns.
Flora	The plants of a particular region and/or period.
Forage	Food for cattle.
Forb	Any herbaceous (non-woody) plant other than grasses or grasslike species.
Fruit	Plant part that contains seeds.
Germinate	In plants, to break seed dormancy and begin to grow or develop.
Girdle	Removing a ring of bark (usually 2–5″ wide) from a tree or branch; often done to kill it.
Habit	The characteristic form or shape of a plant.

Hastate	Leaves with two lobes spreading outward at the base.
Heartwood	The wood in the center of a tree.
Herb (herbaceous)	A flowering plant whose stems die back to the ground during the year, often with a fall frost; tissues are usually not woody like those of trees and shrubs, but roots may continue to produce stems for many years.
Hybrid	The offspring of two closely related species.
Inflorescence	The flower cluster or flower-bearing portion of a plant.
Integrated Vegetation Management (IVM)	Combining two or more basic methods of weed control. This may include prevention, containment, manual/mechanical methods, the use of herbicides, biological controls, and/or the planting of native species for competition.
Invasive plant	A plant that can grow quickly, spread rapidly, and dominate a natural area to the point that other species are displaced, especially those native to an area; economic, environmental, and/or human health factors are at risk if the plant is not controlled.
Irregular (flower)	Not symmetrical; may be lopsided, lipped, etc.
Lanceolate	A leaf shaped like a lance head; longer than it is wide with the widest part below the midpoint of the leaf.
Lateral (branch)	A branch that is to the side of or other than the main stem.
Leaf blade	The wide, free part of a leaf.
Leaflet	One of the sections of a compound leaf.
Leaf scar	The attachment mark left on a twig after a leaf falls off.
Legume	A nitrogen-fixing plant that produces pods containing seeds.
Lenticels	Circular or elongated spots on the surface of bark that function as pores for the exchange of plant gases with the atmosphere.
Ligule	An appendage (a short membrane or row of hairs) on the upper side of the grass leaf where the leaf and blade join.
Linear (leaf)	Long, narrow; veins are parallel.
Lobed (leaf)	Deeply indented, with rounded or pointed outer projections.
Manual	Done by hand or without the use of power equipment.
Margins (leaf)	Edges.
Material Safety Data Sheet (MSDS)	Information outlining and detailing the proper procedures for handling and working with a particular herbicide.
Mattock	A pickax-like tool used to loosen the soil.
Mechanical control	A means of physically removing a species such as burning, mowing, cutting, or girdling.
Membrane	A thin, soft layer of covering tissue on a plant that lets light through without being transparent.
Mesic	The middle of the moisture the continuum from wet to dry.
Midvein	The main vein in the center of a leaf extending from the petiole.

Monocarpic	A plant that produces a single fruit or has only one fruiting period during its life cycle.
Monocot	A plant whose embryo has only one cotyledon; typically grasses or grasslike species, such as sedges, lilies, and orchids.
Monoculture	A plant community comprised of one species.
Monoecious	Having male and female flowers on the same plant.
Monofilament	A threadlike structure, as in a single strand of plastic or wire.
Native plant	A plant known to exist in the upper Midwest prior to European settlement. Also known as an indigenous plant.
Node	The point on the stem from which leaves or branches arise.
Non-native plant	A plant not considered native to a particular area; exotic plant; nonindigenous plant.
Noxious (weeds)	Those plants legally considered harmful to public health, the environment, natural areas, crops, livestock, or other property which, therefore, must be destroyed.
Ovate (leaf)	Egg-shaped leaf with the widest part near the base.
Palatable	Agreeable to the taste.
Palmate (compound leaf)	Divided or lobed so all leaflets radiate from one point (as fingers from a palm).
Panicle (cluster)	An elongated, compound (branched) flower cluster.
Peduncle	A stem supporting a flower cluster or single flower, either leafless or with bracts.
Perennial	A plant that lives more than two growing seasons.
Petiole	A stalk that attaches the leaf blade to the stem.
Photosynthesis	The process by which plant cells make carbohydrates from carbon dioxide and water by using chlorophyll and light.
Pinnate (compound leaf)	Having leaflets or lobes or branches or veins arranged on two sides of a leaf stem or rachis.
Pioneering species	Species that quickly establish themselves in ecologically disturbed communities; also known as "colonizers."
Pistil	The female flower organ that develops fruit; comprised of the stigma, style, and ovary.
Propagules	Plant parts, such as seeds, roots, stems, rhizomes, tubers, and spores, that are capable of producing additional plants.
Pubescent	With soft, downy hairs.
Raceme	A long cluster of flowers or fruit arranged singly along a stalk, each flower with its own small stalk.
Restricted chemicals	Chemicals requiring a license to use.
Rhizome	A rootlike, creeping stem found along or under the ground, and producing roots at nodes.
Rosette	A cluster of leaves surrounding the stem at the base of a plant; for example, basal rosette.
Samara	A simple, dry, one- or two-seeded fruit with winglike outgrowths; typical of maples and other members of the *Acer* genus.
Sapling	A young tree.

Sapwood	The soft, living wood beneath the bark through which sap flows.
Savanna	A plant community with scattered trees and an understory of graminoids and forbs specific to it; a vegetative midpoint on the continuum from prairie to forest.
Scythe	A cutting tool having a long, curved blade fastened at an angle to a handle.
Selective herbicide	An herbicide that is effective against a certain types of plants but does not harm other types of plants.
Sepal	One separate segment of the calyx.
Serrate (leaf)	Toothed, like the edge of a saw.
Sheath (leaf)	Tubular lower portion of a leaf base surrounding its stem.
Shrub	A perennial woody plant, typically growing less than 20′ tall, with several permanent woody stems that tend to be somewhat uniform in diameter.
Simple (leaf)	A leaf with a single blade; undivided, unbranched; as opposed to a compound leaf.
Species	A distinct group of related plants (or animals) with certain common characteristics. They can interbreed to produce offspring.
Spikelet	The basic flowering unit of a grass or sedge.
Stamen	Male flower organ consisting of a slender stalk (filament) and a knoblike tip (anther) that produces pollen. There are usually several stamens per plant.
Stem	The main stalk of a plant.
Stipules	Leaflike appendages, usually paired, at the base of a leafstalk.
Stolons	Horizontal stems on the ground's surface that root at the nodes.
Strain	A variety of a species that has unique characteristics such as color, shape, and size; may be naturally unique or artificially bred.
Subtend	To extend beneath as with a bract; to bear (a flower) in the angle between a leaf and its stem.
Succulent (stems)	Having fleshy and juicy tissues.
Surfactant	A chemical added to an herbicide to reduce the surface tension of water and enhance the ability of the herbicide to be absorbed by the treated plant.
Systemic (herbicide)	Affecting the entire plant.
Taproot	A main root descending downward and giving off small lateral roots.
Taxa	Plant (or animal) classification that includes species, hybrids, subspecies, varieties, forms, and cultivars.
Tendril	A coillike part of a climbing plant that winds around a support.
Terminal (buds)	Situated at the end of stems or twigs.
Translocated species	Species native to a vast area, such as North America, whose range now includes areas where they were not found originally or prior to European settlement.
Tree	A perennial woody plant, typically growing more than 20′ tall, that usually has one main, permanent stem.
Turion	A dormant, vegetative bud.

Umbel	An inverted umbrella-like flower cluster with all flower stalks radiating from the same central point; may be simple or compound.
Understory	Vegetation growing near the ground.
Volatility	The ability of a liquid or solid to change into a gas (vapor) at ordinary temperatures when exposed to air.
Water column	The entire water body profile in cross section, from the surface down to the bottom.
Weed	A plant growing where it is not wanted; a plant that is invasive and persistent in cropland, forests, or natural areas.
Wing	A thin, membranous flap extending along a stem, stalk or other plant part.
Winter annual	A plant that germinates in late summer or fall, overwinters as a low-growing plant, flowers and produces seed the next spring, and then dies.
Xenophobia	Fear or hatred of foreigners or anything strange or foreign.

Matrix of Invasive, Non-native Plants of the Upper Midwest

Growth form	Invasiveness Category	Habitat Invaded	Control Methods	Current Range	Comments
T = Tree	1 = Major invader of natural areas	For = Forests, woodlands	1 = Manual	NA = Widespread in North America	# = Primarily a problem in disturbed areas
S = Shrub	2 = Lesser invader of natural areas	Sav = Savannas, open woodlands, barrens	2 = Mechanical	NE = Northeastern US	~ = Spreads vegetatively, best not to plant adjacent to natural areas
V = Vine	3 a= Likely to become major invader	Gra = grasslands, prairies	3 = Manual or Mechanical followed by Chemical	NW = Northwestern US	* = Hybridizes with native species
F = Forb	3 b= Potential impact unknown	Wet = Wetlands	4 = Chemical	GP = Great Plains	!! = Toxic if ingested
G = Grass or Grass-like (not yet wide-spread in range)	4= Native to all or part of range	Lak = Lakes and/or ponds	5 = Biological	S = Southern US	W = Western US
A = Aquatic		Str = Streams or stream banks	5* = Biological (in development)	UM = Widespread in Upper Midwest	!d = Dermally toxic
		Dun = Dunes or beaches		EM = Eastern part of Midwest	$ = Species used in agriculture, land-scaping, or other beneficial use, but can spread into natural areas
				WM = Western part of Midwest	CV= There are cultivar(s) known or thought to be sterile or non-invasive
				SM = Southern part of Midwest	&= Both native and non-native ecotypes are found in the region
				NM = Northern part of Midwest	
				CA = Southern Canada	

Scientific Name	Common Name	Growth form	Invasiveness category	Habitat invaded	Control Method	Current range	Comments
Acer ginnala	Amur Maple	T	2	For, Gra	1, 2	UM, NE	$ CV
Acer platanoides	Norway Maple	T	2	For	1, 2	NE, NW, NM, EM	$ CV
Aegopodium podagraria	Goutweed	F	2	Gra	1, 4	NE, NW, NM, EM	~ $
Agropyron cristatum	Crested Wheat Grass	G	3b	Gra	1, 2, 4, 5	W, WM, SM	$
Agropyron repens (see *Elytrigia*)	Quack grass						
Agrostis gigantea	Red Top	G	2	Sav, Gra, Wet	1, 2, 4	NA	$ &
Agrostis stolonifera	Creeping Bentgrass	G		Gra	1, 2, 4	NA	
Ailanthus altissima	Tree-of-Heaven	T	2, 3b	Sav, Gra, For	1, 2	NW, NE, EM, SM, S	#
Aira caryophyllea	Silver Hair Grass	G	2	Wet	1, 4	NE, SM, NW, S	
Akebia quinata	Five-Leaf Akebia	V	3b	For, Gra	1	NE, SE, SM	
Albizia julibrissin	Mimosa	T	3b	Gra, For	1	NE, SM	$
Alliaria petiolata	Garlic Mustard	F	1	For, Sav	1, 4, 5*	NE, UM, GP	
Allium vineale	Wild Garlic	F	2	Gra, Wet	1	NE, S, SM, EM, NW	#
Alnus glutinosa	Black Alder	T	2	Wet	1	NE, UM	
Ampelopsis brevipedunculata	Porcelain Berry	V	3a	Gra, Sav	1	NE, EM	$
Anagallis arvensis	Scarlet Pimpernel	F	3b	Wet	1	NA	#
Anchusa officinalis	Common Bugloss	F	2	Gra	1, 4	NW, UW, NE	
Angelica sylvestris	Wood Angelica	F	3b	For, Gra, Sav	1	W	
Anthoxanthum odoratum	Sweet Vernalgrass	G		Gra	1, 4	NE, NW, EM, SM	
Anthriscus sylvestris	Wild Chervil	F	3	For	1, 4	NE, NM, NW	
Arctium minus	Burdock	F	2	For	1	NA	#
Arctium vulgare	Woodland Burdock	F		Gra, Sav	1	NE	
Arthraxon hispidus	Hairy Joint Grass	G	3b	Str, For, Wet	1, 4	NE, S	
Artimesia absinthium	Common Wormwood	F	2	Gra, Sav	1, 4	NW, UM, NE, S	$ &
Belamcanda chinensis	Blackberry Lily	F	3b	For, Gra, Wet	1	NE, S, UM	
Berberis thunbergii	Japanese Barberry	S	2	For, Wet, Sav	1, 4	NE, GP, UM	$ CV
Berberis vulgaris	European or Common Barberry	S	2	Gra, Sav	1, 4	NE, NW, UM	$ CV
Berteroa incana	Hoary Allysum	F	2	Gra	1, 4	W, NW, UM, NE	# !i
Betula pendula	European Weeping Birch	T	3b	Gra, Wet	1	NE, EM, NW	$
Bothriochloa bladhii syn. *Andropogeton caucasicus*	Caucasian Bluestem	G	3b	Gra, Sav	1, 2, 4	GP, SS	$
Bromus inermis	Smooth Brome	G	2	Sav, Gra	1, 2, 4	NA	$ &

Scientific Name	Common Name	Growth form	Invasiveness category	Habitat invaded	Control Method	Current range	Comments
Bromus tectorum	Cheat Grass	G	3b	Gra	1, 2, 4	NA	#
Bromus japonicus	Japanese Brome	G		Gra	1, 2, 4	NA	
Butomus umbellatus	Flowering Rush	A, F	3b	Lak, Str	1	NE, NM, GP	
Cabomba caroliniana	Fanwort	A, F	3b	Lak, Str	1, 4	S, NE, NW, SM	$
Campanula rapunculoides	Creeping Bellflower	F	2	For	1	NA	#
Caragana arborescens	Siberian peashrub	S	2	Gra, Sav	1	NM, UM	$
Carduus acanthoides	Plumeless Thistle	F	2	Gra	1, 4	NE, GP, NM, NW	#
Carduus nutans	Musk Thistle	F	2	Gra	1, 4, 5	NA	#
Celastrus orbiculatus	Round-Leaved or Oriental Bittersweet	V	1	For, Gra, Sav	1, 3, 4	NE, UM	$
Centaurea biebersteinii, syn. C. maculosa	Spotted Knapweed	F	1	Gra, Dun, Sav	1, 4, 5	NA	# !i !d
Centaurea solstitialis	Yellow Starthistle	F	3	Gra	1, 4, 5	NA	!i
Centaurea cyanus	Bachelor's Button	F	2	Gra	1, 4	NA	$
Centaurea diffusa	White-Flowered or Diffuse Knapweed	F	2	Gra	1, 4	W, EM, SM	
Centaurea maculosa (see C. biebersteinii)							
Centaurea repens syn. Acroptilon repens	Russian Knapweed	F	2	Gra	1, 4	W, UM	
Chelidonium majus	Celandine	F	2	For	1, 4	NE, UM, NW	#
Chrysanthemum leucanthemum	Ox-Eye Daisy	F	2	For, Gra	1, 4	NA	# $
Cichorium intybus	Chicory	F	2	Gra	1, 2, 4	NA	#
Cirsium arvense	Canada Thistle	F	1	For, Gra	1, 2, 3, 4, 5	NA	#
Cirsium palustre	European Marsh Thistle	F	3a	Wet, For	1, 4	NM, NE	
Cirsium vulgare	Bull Thistle	F	2	For	1, 2, 4	NA	#
Clematis terniflora	Sweet Autumn Leatherleaf Clematis	V	2	For, Sav	1	SM, SE	$
Conium maculatum	Poison Hemlock	F	2	Wet	1, 4	NA	!i
Convallaria majalis	Lily-of-the-Valley	F	2	For	1, 4	NE, UM, NW	!i ~ $
Convolvulus arvensis	Field Bindweed, Creeping Jenny	F, V	2	Str, Gra	1, 4	NA	!i ~
Coronilla varia	Crown Vetch	F, V	1	Gra, For, Str	1, 4	NA	# ~ $
Crateagus monogyna	Single-seed Hawthorn	T, S	2	Gra, Sav	1	UM, NE	
Crepis tectorum	Hawksbeard	F	2	Gra	4	UM, GP, NE, NW	

Scientific Name	Common Name	Growth form	Invasiveness category	Habitat invaded	Control Method	Current range	Comments
Cynanchum spp. (see *Vincetoxicum*)							
Cynoglossum officinale	Hound's Tongue	F	2	Gra	1	NA	#
Cytisus scoparius	Scotchbroom	S	3b	Gra, Sav	1, 4	GP, NW, NE, EM, S	
Dactylis glomerata	Orchardgrass	G	2	Gra	1, 4	NA	$
Daucus carota	Queen Anne's Lace, Wild Carrot	F	2	For, Gra	1, 2, 4	NA	#
Dianthus armeria	Deptford Pink	F	2	Gra	1	NE	
Digitalis lanata	Grecian Foxglove	F	3b	Gra	1, 4	NE, NM, EM	!i $
Digitalis lutea	Yellow Foxglove	F	3b	For, Gra	1, 4	NM, NE	$
Digitalis purpurea	Foxglove	F	2	Gra, For	1, 4	NW, NE, NM	!i
Dioscorea oppositifolia	Chinese Yam	V	3b	For	1, 3, 4	S, SM, NE	
Dipsacus laciniatus	Cut-Leaved Teasel	F	1	Gra, Sav, Wet	1, 2, 4	NE, UM	$
Dipsacus sylvestris	Common Teasel	F	1	Gra, Sav, Wet	1, 2, 4	EM, SM, W	$
Echinochloa crusgalli	Barnyard Grass	G	2	Wet	1, 2, 4	NA	
Echinops sphaerocephalus	Globe Thistle	F	3b	Gra	1, 2, 4	SM, NE	$
Egeria densa	Brazillian Waterweed	A, F	3b	Lak, Str	1, 2, 4	SE, NE, W	
Eichhornia crassipes	Water Hyacinth	A, F	3b	Lak, Str	1, 2, 4	S, NE, W	$
Elaeagnus angustifolia	Russian Olive	T	2	Gra, Str	1	NE, NW, GP, UM	$
Elaeagnus umbellata	Autumn Olive	S	1	Gra, Sav, For	3	NE, S, SM, EM	# $
Elsholtzia ciliata	Elsholtzia	F	3b	Gra	4	NE, NM	
Elytrigia repens	Quack Grass	G	2	Sav, Dun	1, 2, 4	NA	# $
Epilobium hirsutum	Hairy Willow-Herb	F	3b	Wet	1	NE, NM, EM, NW	
Epipactis helleborine	Helleborine	F	2	For	1	NE, UM, W	~
Euonymus alata	Burning Bush, Winged Euonymus	S	2	For, Gra	1	NE, NM	$ CV
Euonymus fortunei	Winter Creeper	V	2	For	1, 3, 4	EM, SM, NE	~ $
Euphorbia cyparissias	Cypress Spurge	F	2	For, Gra, Sav	1, 4	NW, GP, UM, NE	!i !d ~
Euphorbia esula	Leafy Spurge	F	1	Gra, Sav	1, 4, 5	NW, GP, UM, NE	!i
Fallopia (see *Polygonum*)							
Festuca arundinacea syn. *Festuca elatior*	Tall Fescue	G	2	Gra, Sav, Wet	1, 2, 4	NA	$!i
Festuca pratensis	Meadow Fescue, Bedstraw	G	3b	Gra	1, 4	NA	$
Filipendula ulmaria	Queen-of-the-Meadow	F	3b	Wet, For	1	NM, NE	
Galeopsis tetrahit	Hemp Nettle	F	2	For, Dun	1	UM, NE, NW, GP	#

Scientific Name	Common Name	Growth form	Invasiveness category	Habitat invaded	Control Method	Current range	Comments
Galium odoratum	Sweet Woodruff	F, V	2	For, Sav	1, 4	NW, WM, NE	$
Geranium ibericum	Iberian Geranium	F	3b	For, Gra	1	NE	
Geum urbanum	Avens, Herb-Bennet	F	2	For	1	NM, SM, NE, NW	*
Glaucium flavum	Yellow Horn Poppy	F	3b	Lak, Dun	1	NE, EM, W	
Glechoma hederacea	Creeping Charlie, Creeping Jenny	F	2	For	1, 4	NA	~
Glyceria maxima	Tall Mannagrass	G	3b	Wet	1	EM	
Gypsophila paniculata	Baby's Breath	F	2	Dun	1	NE, NW, EM, NM	$
Hedera helix	English Ivy	V	2	For	1, 3, 4	S, SM, NE	
Heliotropium indicum	Indian heliotrope	F	2	Wet	1	SM, S, NE	#
Hemerocallis fulva	Orange Day Lily	F	2	For, Gra	1, 2, 4	NE, S, NW, UM	$ ~
Heracleum mantegazzianum	Giant Hogweed	F	3a	Wet, Str	1, 2, 4	NE, NW, EM	!d
Hesperis matronalis	Dame's Rocket	F	2	Gra, Sav, For, Wet	1, 2, 4	NE, NW, UM, GP	$
Hibiscus syriacus	Rose of Sharon	S	3b	Gra	1, 4	S, SM, NE	$
Hieracium aurantiacum	Orange Hawkweed,	F	2	For, Sav, Gra, Dun	4	NE, EM, NM, NW	
Hieracium caespitosum syn. *H. pratense*	Devil's Paintbrush Yellow Hawkweed, King Devil	F	2	Gra, For	4	NE, UM, NW	
Hieracium lachenalii	Hawkweed	F	2	Gra, Sav	4	NE, NM, NW	
Hieracium piloselloides	Yellow or Tall Hawkweed	F	2	Gra, Sav	4	NE, UM	
Humulus japonicus	Japanese Hops	V	3b	Gra	1	SM, NE, EM	$
Hydrilla verticillata	Hydrilla	A, F	3a	Lak, Str	1	S, W, NW	
Hydrocharis morsus-ranae	European Frog-bit	A, F	3a	Lak, Str, Wet	1, 2	NE	$
Hypericum perforatum	Common St. John's Wort	F	2	For, Dun	1, 3, 4	NA	$!i
Impatiens glandulifera	Ornamental Jewelweed	F	3b	Str, For, Wet	1	NE, EM, NW	
Inula helenium	Elecampane Inula	F	2	Gra, For, Wet	1, 4	NM	
Iris flavenscens	Lemonyellow Iris	F	3b	Wet	1	SM	!i
Iris pseudacorus	Yellow (Water-Flag) Iris	F, A	2	Wet, Lak, Str	1	NE, NW, UM, S	$!i
Isatis tinctoria	Dyer's Woad	F	3b	Gra	2, 4	GP, NE, W, SM	
Juncus ensifolius	Sword-leaf Rush	A, G	2	Wet	1	NW, NM, W	
Kochia scoparia	Summer Cypress, Mexican Fireweed	F	2	Gra, Str	4	NA	#
Kummerowia stipulaceae	Korean Clover	F	3b	Gra	4	S, SM, EM, NE	$
Kummerowia striata	Japanese Clover	F	3b	Gra	4	S, SM, NE	$
Lapsana communis	Nipplewort	F	2	For	1	NE, UM, NW	

Scientific Name	Common Name	Growth form	Invasiveness category	Habitat invaded	Control Method	Current range	Comments
Lathyrus latifolius, L. sylvestris	Everlasting Pea	V, F	2	Gra	1, 4	NA	$
Leonurus cardiaca	Motherwort	F	2	For	1	NA	#
Lepidium latifolium	Pepperweed	F	2	Wet, Dun, Str	1	NW, W, SM, NE	
Lespedeza bicolor	Shrub Lespedeza	S	2, 3b	Gra	2, 4	NE, EM, S, SM	$
Lespedeza cuneata syn. *Lespedeza sericea*	Chinese Lespedeza, Sericea Bush Clover	F	3a	Sav, Gra	2, 4	SM, EM, NE, S	$
Lespedeza stipulaceae (see Kummerowia stipulaceae)							
Lespedeza thunbergii	Shrubby Bush Clover	F	3b	Gra	4	UM, SM, SE	
Ligustrum obtusifolium	Blunt-leaved Privet	S	2	For, Gra	3, 4	NE, EM, SM	$
Ligustrum ovalifolium	California Privet	S	2	Wet, For	3, 4	EM, S, W	$
Ligustrum sinense	Chinese Privet	S	2	For	3, 4	S	$
Ligustrum vulgare	Common Privet	S	2	Wet, For, Gra	3, 4	NE, EM, S, NW	$
Linaria dalmatica	Dalmation Toadflax	F	2	Gra	4, 5	GP, NW, UM, W, NE	
Linaria vulgaris	Butter-and-Eggs	F	2	Gra	1, 4	NA	#
Lonicera fragrantissima	Winter Honeysuckle	S	3b	For	1	SE, NE, EM	$
Lonicera japonica	Japanese Honeysuckle	V	2, 3a	For, Sav, Gra	1, 3, 4	S, SM, NE, EM, W	$
Lonicera maackii	Amur Honeysuckle	S	1, 3a	For, Sav, Gra	1, 3, 4	NE, EM, SM	$
Lonicera morrowii	Morrow's Honeysuckle	S	1	Gra, Sav, For	1, 3, 4	NE, NM, EM	$
Lonicera tatarica	Tartarian Honeysuckle	S	1	Gra, Sav, For	1, 3, 4	UM, NE, GP, W	$
Lonicera x bella	Bell's Honeysuckle	S	1	Gra, For	1, 3, 4	NE, UM	$
Lonicera xylosteum	Dwarf Honeysuckle	S	2	For, Sav, Gra	1, 3, 4	NE, SM, NW, EM	$
Lotus corniculata	Bird's-foot Trefoil	F, V	1	Gra	1, 2, 4	NA	$ ~
Lupinus polyphyllus	Big-leaf Lupine	F	2	Gra, Sav	1, 2, 4	NM, NE, NW	$!i
Lychnis alba	Bladder Campion	F	2	Gra	1, 4	UM, NE, NW, GP	#
Lycopus europaeus	European Water Horehound	A	3a	Lak, Str, Wet	1	NE, EM	
Lysimachia nummularia	Moneywort	F, V	2	Wet, For	1, 4	UM, NE, S, NW	$ ~
Lysimachia vulgaris	Yellow Garden Loostrife	F	2	Wet, Gra	1, 2, 4	NE, EM, SM, NW	$
Lythrum salicaria	Purple Loosestrife	F, A	1	Wet	1, 3, 4, 5	NA	$
Lythrum virgatum	European Whorled Loostrife	F	2	Gra, Wet	1, 2, 4	NE	#
Maclura pomifera	Osage Orange	T, S	4	For	1, 2, 3	S, SM, NE, NW, EM	
Marisilea quadrifolia	Water Shamrock	A, F	3b	Lak, Str	1, 4	NE, SM	$
Medicago lupulina	Black Medick	F	2	Gra	1, 2, 4	NA	

Scientific Name	Common Name	Growth form	Invasiveness category	Habitat invaded	Control Method	Current range	Comments
Melilotus alba	White Sweet Clover	F	1	Gra, Sav	1, 2, 4	NA	$
Melilotus officinalis	Yellow Sweet Clover	F	1	Gra, Sav	1, 2, 4	NA	$
Mentha spicata	Spearmint	F	2	Wet	1, 4	NA	$
Mentha x piperita	Peppermint	F	2	Wet	1, 4	NA	$
Microstegium vimineum	Japanese Stilt Grass	G	3a	For, Sav, Wet	1, 4	SM, S, NE	
Miscanthus sinensis	Chinese Silver Grass, Eulalia	G	2	Gra	1, 4	S, NE, EM	$ ~
Morus alba	White Mulberry	T	2	For	1, 3	NA	#
Myosotis sylvatica	Garden or Woodland Forget-Me-Not	F	2	For, Wet	1, 4	NE, UM, NW	
Myosotis scorpioides	Common or Water Forget-Me-Not	F, A	2	Wet, Lak, Str	1, 4	NE, UM, NW, W, S	~
Myriophyllum aquaticum	Parrot Feather	A, F	3a	Lak, Str	1, 2, 4	S, NW, W, NE	$
Myriophyllum heterophyllum	Twoleaf Water-Milfoil	A, F	3b	Lak, Str	1, 2, 4	S, GP, NE, UM	
Myriophyllum spicatum	Eurasian Water-Milfoil	A, F	1	Lak, Str	1, 2, 4, 5	S, NE, UM, NW, W	
Najas minor	Lesser Naiad	A, F	3b	Lak	1, 2, 4	S, NE, EM	
Nymphoides peltata	Yellow Floatingheart	A, F	3b	Lak, Str	1	NE, SM, S, NW, W	$
Onoperdum acanthium	Scotch Thistle	F	3b	Gra	1, 4	GP, NW	
Ornithogalum umbellatum	Star of Bethlehem	F	2	For	1, 4	NE, UM, S, NW	$!i
Pachysandra terminalis	Pachysandra	F	2	For	1, 4	NE, NM, SM	$~
Pastinaca sativa	Wild Parsnip	F	1	Gra, Sav	1, 2, 4	NA	!d $
Paulownia tomentosa	Princess Tree	T	3a	For, Str	1, 3, 4	NE, SM, S	$
Perilla frutescens	Beefsteak Plant	F	2	Gra, For	1, 4	NE, S, UM	#
Phalaris arundinacea	Reed Canary Grass	G, A	1	Wet, Str, Gra, For	1, 2, 3, 4	NA	& $ *
Phalaris canariensis	Canary Grass	G	2	Gra, Wet	1, 2, 3, 4	NA	
Phellodendron amurense	Japanese Cork Tree	T	3b	For, Wet	1, 3, 4	NE, SM	$
Phleum pratense	Timothy	G	2	Gra	2, 4	NA	$
Phragmites australis	Giant Reed Grass, Phragmites	G, A	1	Wet, Str	3, 4	NA	& $ *
Pimpinella saxifraga	Burnett Saxifrage	F	3b	Wet	1	NE, NM, EM, NW	
Pinus sylvestris	Scotch Pine	T	2	Gra, Sav	1	NE, UM	$
Pistia stratiotes	Water Lettuce	A, F	3b	Str, Lak	1, 4	S, NE, W	$
Poa bulbosa	Bulbous Bluegrass	G	2	Gra, Sav	1	NW, GP, EM, SM	
Poa compressa	Canada Bluegrass	G	2	Gra, Sav	1	NA	$
Poa pratensis	Kentucky Bluegrass	G	2	Gra, Sav	1	NA	$

Scientific Name	Common Name	Growth form	Invasiveness category	Habitat invaded	Control Method	Current range	Comments
Poa trivialis	Rough Bluegrass	G	2	Gra, Sav, Wet	1	NW, NE, UM	
Polygonum convolvulus	Black Bindweed or	V, F	2	For	1	NA	
Polygonum cuspidatum syn. *Fallopia japonica*	Wild Buckwheat Japanese Knotweed	F	1	For, Str, Dun	1, 2, 3, 4	EM, NE, NW	~
Polygonum sachalinense syn. *Fallopia sachalinensis*	Giant Knotweed	F	3b	Str, For	1, 2, 3, 4	NE, NW, EM	~
Polygonum perfoliatum	Mile-A-Minute Vine	V, F	3b	For, Gra, Wet	1, 4	NE, NW	
Populus alba	White Poplar	T	2	Gra	1, 3	NA	$
Populus nigra	Lombardy Poplar	T	2	Dun	1, 3	NE, UM, S	$
Potamogeton crispus	Curly-Leaf Pondweed	A, F	2	Lak, Str	1	NA	
Potentilla argentea	Silver Cinquefoil	F	2	Gra	1, 4	NE, UM, NW, GP	
Potentilla recta	Sulphur Cinquefoil	F	3b	Gra	1	NA	
Prunella vulgaris	Heal-all, Self-heal	F	2	Gra	1	NA	
Prunus avium	Sweet Cherry	T	3b	For	1	NE, EM, NW	$
Prunus mahaleb	Mahaleb Perfumed Cherry	T, S	3b	Gra	1	NA	
Pueraria lobata	Kudzu	V	3b	For, Sav, Gra	1, 2, 3, 4	S, SM, NE	$
Pyrus calleryana	Callery Pear	T	3b	Gra	1	S, SM	$
Quercus acutissima	Sawtooth Oak	T	3b	For	1	S, NE	$
Ranunculus acris	Tall buttercup	F	2	Gra	1, 4	NE, UM, NW, GP	$ & !i
Ranunculus ficaria	Lesser Celandine	F	2	Gra, For	1, 4	NE, EM, NW	~ !i
Ranunculus repens	Creeping buttercup	F	2	For, Gra, Wet	1, 4	NE, UM, NW, GP, W	~ $!i
Rhamnus cathartica	Common Buckthorn	T, S	1	For, Sav, Wet	1, 3, 4	NE, UM, GP, W	$!! CV
Rhamnus davurica	Dahurian buckthorn	S, T	3b	For, Sav	1, 3, 4	EM, SM	!i
Rhamnus frangula syn. *Frangula alnus*	Alder, Glossy or Columnar Buckthorn	T, S	1	Wet, For	1, 3, 4	UM, NE	$ CV
Rhodotypos scandens	Jet-bead	S	2	For	1, 4	SM, NE, EM	$
Robinia hispida	Bristly Locust	S	2	Gra, Sav	1, 3, 4	NE, UM	$
Robinia pseudoacacia	Black Locust	T	1	For, Gra	1, 3, 4	NA	$!i
Robinia viscosa	Clammy Locust	S	3b	Gra, Sav	1, 3, 4	NE, SM	
Rorippa nasturtium-aquaticum	Watercress, Yellowcress	A, F	2	Lak, Str	1	NA	$ ~
Rosa multiflora	Multiflora Rose	S	1	Gra, Sav, For	1, 3, 4, 5	UM, NE, S, NW	$
Rosa rugosa	Rugosa Rose	S	2	Gra	1, 3, 4	NE, NM, EM, NW	$
Rubus phoenicolasius	Wineberry	S	3b	Gra	2, 3, 4	NE, EM, SM	

Scientific Name	Common Name	Growth form	Invasiveness category	Habitat invaded	Control Method	Current range	Comments
Rumex acetosella	Field, Red, or Sheep Sorrel	F	2	Gra	2, 4	NA	#
Rumex crispus	Curly dock	F	2	Gra, Sav, Wet	1	NA	
Salix alba	White Willow	T	2	Wet	1, 2, 3	UM, NE, NW, W, GP	*
Salix fragilis	Crack Willow	T	2	Wet	1, 2, 3	UM, NE, NW, GP	*!i
Salsola tragus	Russian Thistle, Tumbleweed	F	3b	Gra		W, WM	
Saponaria officinalis	Bouncing Bet or Soapwort	F	2	Gra	1	NA	
Sedum acre	Yellow Sedum	F	2	Gra	1	NE, UM, NW	$
Sedum purpureum	Live Forever, Purple Sedum	F	2	Gra, Wet	1	NE, UM, NW	$
Sedum sarmentosum	String Stonecrop	F	3b	Gra	1	NE, EM, SM	$
Senecio jacobaea	Tansy Ragwort	F	3b	Gra	1, 4, 5	NW, EM, NE	!i #
Setaria faberi	Giant Foxtail	G	2	Gra	1, 2, 4	NA	
Setaria viridis	Green Foxtail	G	2	Gra	1, 2, 4	NA	
Silybum marianum	Milk Thistle	F	2	Gra	1	NE, EM, NW, W, S	$ # !i
Silene latifolia	White Campion	F	2	Gra	1, 4	UM, NE, NW, GP	#
Silene vulgaris	Bladder Campion	F	2	Gra	1, 4	NE, UM, GP, NW	
Solanum dulcamara	Deadly or Bittersweet Nightshade	V, F	2	For	1	UM, NE, NW, GP	# !i
Sonchus arvensis	Perennial or Field Sowthistle	F	2	Gra, Str	1, 2	UM	# !i
Sorbus aucuparia	European Mountain-Ash	T	2	For	1	NE, UM, NW	$
Sorghum halepense	Johnson Grass	G	2	Gra, Wet	1, 4	NA	!i
Spiraea japonica	Japanese Spiraea or Japanese Meadowsweet	S	2	For, Gra, Str	1	NE, EM, SM	$ #
Stellaria aquatica	Giant Chickweed	F	2	Wet	1	NE, UM	
Stellaria media	Common Chickweed	F	2	For, Wet	1	NA	#
Tamarix ramosissima	Salt Cedar	T, S	3a	Gra, Str	1, 2, 3	W, S, WM	$
Tanacetum vulgare	Tansy	F	2	Gra, Sav	1, 4	NA	$!i
Torilis arvensis	Field Hedgeparsley	F	3b	Gra, Sav, For	1, 4	S, SM, NW, W	
Torilis japonica	Japanese Hedgeparsley	F	3b	Gra, Sav, For	1, 4	EM, NE, S, NW	
Trapa natans	Water Chestnut	A, F	3a	Lak, Str	1	NE	$
Trifolium hybridum	Alsike Clover	F	2	Gra	2, 4	NA	$ #
Trifolium pratense	Red Clover	F	2	Gra	4	NA	$ #
Trifolium repens	White Clover	F	2	Gra	4	NA	$ #
Tussilago farfara	Colt's-foot	F	2	Gra, Wet	1	NE, UM	
Typha angustifolia	Narrow-leaved Cattail	G	1	Wet Str	1, 2, 3, 4	UM, NE, NW, W, GP	$ *

Scientific Name	Common Name	Growth form	Invasiveness category	Habitat invaded	Control Method	Current range	Comments
Typha x glauca	Hybrid Cattail	G	1	Wet, Str	1, 2, 3, 4	NE, UM, W	*
Ulmus pumila	Siberian Elm	T	2	Wet, For	1, 3	NA	
Urtica dioica	Stinging Nettle	F	2	Wet	1, 2	NA	* &
Valeriana officinalis	Garden Heliotrope	F	2	Gra, For, Wet	1	UM, NE, NW	$
Verbascum thapsus	Mullein	F	2	Gra	1, 2	NA	#
Verbena bonariensis	Purple Top Vervain	F	3b	Gra	1	S, NE, NW	$
Viburnum lantana	Wayfaring Tree	S	2	For	1, 3	NE, EM	$
Viburnum opulus subsp. *opulus*	European Highbush Cranberry	S	2	For	1, 3	NE, UM, GP, NW	$ &
Vinca minor	Periwinkle	V	2	For, Sav	1, 4	UM, NE, S, NW	$ ~
Vicia cracca	Cow Vetch, Bird Vetch	F, V	2	Gra	1, 2, 4	UM, NE, NW, W	$
Vicia villosa	Hairy Vetch, Winter Vetch	F, V	2	Gra	1, 2, 4	NA	$
Vincetoxicum hirundinaria	Swallow-Wort	V	3b	For	1, 4	NE	
Vincetoxicum nigrum syn. *Cyanchum nigrum, C. louiseae*	Black Swallow-Wort, Dog-Strangling Vine	V	3a	Gra, Sav, For	1, 4	NE, SM, EM, W	~
Vincetoxicum rossicum syn. *Cyanchum rossicum*	Dog-Strangling Vine	V	3a	Gra, Sav, For	1, 4	NE, SM, EM	~
Wisteria sinensis	Chinese wisteria	V	3b	For	3, 4	S, NE, SM, EM	$ ~

Disclaimer: Listed above are the species for which there is information about invasiveness and spread in the Midwest, regardless of control potential. Confirmation or details may not be available. This list is not comprehensive. The absence of a species on this list does not preclude it from being invasive. This list has no legal bearing or regulatory implications. It is to be used for information only.

Note: This matrix was compiled with the assistance of Kelly Kearns, Rachel Armstrong, and Matthew Burczyk, Wisconsin Department of Natural Resources. Revisions or additions to the matrix are welcome. Please send them to kelly.kearns@dnr.state.wi.us.

References

PRINT MATERIAL

Applied Biochemists, Inc. 1976. *How to Identify and Control Water Weeds and Algae.*

Beule, John D. 1979. *Control and Management of Cattails in Southeastern Wisconsin Wetlands.* Wisconsin Department of Natural Resources, Technical Bulletin No. 112.

Borman, Susan, Robert Korth, and Jo Temte, Wisconsin Lakes Partnership. 1997. *Through the Looking Glass, a Field Guide to Aquatic Plants.* Merrill, WI: Reindl Printing, Inc.

Bosworth, Sid. 2000. *Wild Chervil—A Relatively New Weed Problem in Central Vermont.* University of Vermont Extension in cooperation with the U.S. Department of Agriculture.

Chicago Botanic Garden. 2002. Chicago Botanic Garden Invasive Plant Policy.

City of Boulder Open Space Department. *Warning: Danger in the Field.*

County Weed Directors Association of Kansas, Kansas State Board of Agriculture, and Kansas State University. *Sericea Lespedeza.* Brochure.

Courtenay, Booth, and James H. Zimmerman. 1978. *Wildflowers and Weeds.* New York: Simon and Schuster.

Craven, Scott R., Phillip J. Pellitteri, and Robert C. Neuman. 1993. *Outdoor Hazards in Wisconsin: A Guide to Noxious Insects, Plants, and Wildlife.* University of Wisconsin Cooperative Extension Publication No. 3573.

Dirr, Michael A. 1997. *Dirr's Hardy Trees and Shrubs.* Portland, OR: Timber Press, Inc., 1997.

Eagan, David J. 1999. Burned by Wild Parsnip. *Wisconsin Natural Resources.* June.

———. 2000. Wild Parsnip II. *Wisconsin Natural Resources.* June.

Eggers, Steve. 2001. *Ecology and Control of Phragmites.* St. Paul, MN: U.S. Army Corps of Engineers.

Ellis, Mel. 1996. *Notes from Little Lakes.* Mishicott, WI: The Cabin Bookshelf.

Glass, Steve. 1998. *The Weed Warrior's Handbook.* Blooming Grove, WI: Prairie Fire Press.

Gleason, Henry, and Arthur Cronquist. 1991. *Manual of Vascular Plants of Northeastern United States and Adjacent Canada.* 2nd ed. Bronx, NY: New York Botanical Garden.

Haragan, P.D. 1991. *Invasive Weeds of Kentucky and Adjacent States: A Field Guide.* Lexington: University of Kentucky Press.

Henderson, Carrol L. 1987. *Landscaping for Wildlife.* St. Paul: Minnesota Department of Natural Resources.

Henderson, Rich. 2001. Natural Range Limits: Don't Worry, Be Happy? *The Prairie Promoter* 14(4).

Hightshoe, Gary L. 1988. *Native Trees, Shrubs, and Vines for Urban and Rural America: A Planting Design Manual for Environmental Designers.* New York: Van Nostrand Reinhold.

Hoffman, Randy and Kelly Kearns, editors. 1998. *Wisconsin Manual of Control Recommendations for Ecologically Invasive Plants.* Madison: Wisconsin Department of Natural Resources, Bureau of Endangered Resources.

Kearns, Kelly. 2003. *Garlic Mustard Control: The Latest Thoughts from Kelly Kearns.* Madison: Wisconsin Department of Natural Resources.

Knobel, Edward. 1980. *Field Guide to the Grasses, Sedges and Rushes of the United States.* New York: Dover Publications, Inc.

Lawrence, Eleanor, and Cecilia Fitzsimons. 1985. *An Instant Guide to Trees.* Stamford, CT: Longmeadow Press.

Martin, Alexander C., Herbert S. Zim, and Arnold L. Nelson. 1951. *American Wildlife & Plants—A Guide to Wildlife Food Habits.* New York: Dover Publications, Inc.

Martin, Tunyalee, Wildland Invasive Species Team. 2002. *Weed Alert!: Glyceria maxima.* The Nature Conservancy.

Mello, Kim. 2001. "Leafy Spurge and Cypress Spurge." Presentation at the Plants Out of Place Conference, Eau Claire, WI.

Mills, G. Scott, John B. Dunning, Jr., and John M. Bates. 1989. Effects of Urbanization on Breeding Bird Community Structure in Southwestern Desert Habitats. *The Condor,* 91:416–428.

Minnesota Department of Natural Resources. 2002. *Minnesota Invasive Non-native Terrestrial Plants: An Identification Guide for Resource Managers.*

Mortensen, Carol Estes. 2000. *Weeds of the Northern Lake States.* United States Forest Service and Leech Lake Reservation.

———. 2002. *Invasive Plants of the Southern Tier.* United States Department of Agriculture, Forest Service Region 9.

———. N.d. Is It a Wildflower, or Is It a Weed? Leech Lake Reservation Division of Resources Management and the Chippewa National Forest.

National Audubon Society. 1997. *Familiar Trees of North America—East.* New York: Chanticleer Press, Inc.

Newcomb, Lawrence. 1977. *Newcomb's Wildflower Guide.* Boston, New York, Toronto, London: Little, Brown and Company.

Niering, William A., and Nancy C. Olmstead. 1995. *National Audubon Society Field Guide to North American Wildflowers.* New York: Alfred A. Knopf, Inc.

Ohio Department of Natural Resources, Division of Natural Areas and Preserves and The Nature Conservancy, Ohio Chapter. 1999. *Fighting Invasive Non-Native Plants In Ohio.*

Ohio Department of Natural Resources, Division of Natural Areas and Preserves, The Nature Conservancy, Ohio Chapter, and Columbus and Franklin County Metro Parks. 2000. *Invasive Plants of Ohio.*

Packard, Stephen, and Cornelia Mutel, editors. 2001. *The Tallgrass Restoration Handbook.* Washington, DC: Island Press.

Pauly, Wayne R. 1988. *How to Manage Small Prairie Fires.* Dane County Park Commission, Madison, WI.

———. 2001. *Some thoughts on Glyphosate (RoundUp) Stump Treatment.* Dane County Park Commission, Madison, WI.

Peterson, Roger Tory, and Margaret McKenny. 1968. *A Field Guide to Wildflowers, Northeastern and North-Central North America.* Boston, MA: Houghton Mifflin Company.

Petrides, George A. 1986. *A Field Guide to Trees and Shrubs.* Boston, New York: Houghton Mifflin Company.

Piaskowski, Victoria D. and Gene Albanese. 2001. "Resource sampling of arthropods in all vegetation strata and correlation with arthropods identified in fecal samples of insectivorous warblers at a spring migration stopover site." Presentation at the American Ornithologists' Union Meeting, Seattle, WA.

Powers, Randy. 2000. *Weed Control in New Native Meadows and Gardens.* Prairie Future Seed Co., LLC.

Prairie Ridge Nursery. 2002. *Wildflowers and Native Grasses, 2002 Catalog and Growing Guide.* Mt. Horeb, WI.

Randall, John and Janet Marinelli, editors. 1996. *Invasive Plants: Weeds of the Global Garden.* Brooklyn, NY: Brooklyn Botanic Garden.

Reinartz, Jim, editor. 2003. Invasive Plant Association of Wisconsin's Working List of the Invasive Plants of Wisconsin.

Royer, France and Richard Dickinson. 1999. *Weeds of the Northern U.S. and Canada.* Edmonton: The University of Alberta Press and Renton, WA: Lone Pine Publishing.

Smith, Tim, editor. 1993. *Vegetation Management Manual.* Jefferson City: Missouri Department of Conservation.

Spellenberg, Richard. 1986. *National Audubon Society Pocket Guide, East, Familiar Flowers of North America.* New York: Alfred A. Knopf, Inc.

Symonds, George W. D. 1963. *The Shrub Identification Book.* New York: William Morrow & Company.

Thieret, J. W., and N. C. Olmstead. 2001. *National Audubon Society Field Guide to North American Wildflowers-Eastern Region.* New York: Random House.

Tu, M. 2000. *Techniques from TNC stewards for the eradication of Lythrum salicaria (purple loosestrife) and Phragmites australis (common reed/Phragmites) in wetlands.* The Nature Conservancy.

United States Department of Agriculture, Agricultural Research Service. 1971. *Common Weeds of the United States.* New York: Dover Publications, Inc.

United States Department of Agriculture, Animal and Plant Health Inspection Service. 1992. *Biological Control of Leafy Spurge.* Washington, DC: United States Department of Agriculture.

Uva, Richard H., Joseph C. Neal, and Joseph M. Ditomaso. 1997. *Weeds of the Northeast.* Ithaca, NY: Cornell University Press.

Vance, F. R., J. R. Jowsey, J.S. McLean, and F.A. Switzer. 1999. *Wildflowers of the Northern Great Plains.* Minneapolis: University of Minnesota Press.

Voss, Edward G. 1972. *Flora of Michigan.* Vols. 1–3. Bloomfield Hills, Michigan: Cranbrook Institute of Science.

Washington, Joe. 2001. "Spotted Knapweed and Canada Thistle." Presentation at the Plants Out of Place Conference, Eau Claire, WI..

Watts, May Theilgaard. 1963. *Master Tree Finder.* Rochester, NY: Nature Study Guide Publishers.

Weiss, Laurie. Some Plants Can Be Unfriendly to Humans, Animals. 1998. *Community Newspapers, Inc.,* July 30.

Welsch, Jeff. *Guide to Wisconsin Aquatic Plants.* WIPUBL-WR-173 92 rev. Madison: Wisconsin Department of Natural Resources.

Westbrooks, Randy. 1998. *Invasive Plants—Changing the Landscape of America: Fact Book.* Washington, DC: Federal Interagency Committee for the Management of Noxious and Exotic Weeds (FICMNEW).

Wetter, Mark Allen, Theodore S. Cochrane, Merel R. Black, Hugh H. Iltis, and Paul E. Berry. 2001. *Checklist of the Vascular Plants of Wisconsin.* Technical Bulletin No. 192. Wisconsin Department of Natural Resources in cooperation with the University of Wisconsin-Madison Herbarium, Department of Botany.

Whitson, Tom. 2000. *Weeds of the West.* Laramie, WY: Western Society of Weed Science.

Wild Ones Local Ecotype Committee. 2002. Guidelines for Selecting Native Plants: The Importance of Local Ecotype. *Wild Ones Natural Landscapers Journal* 15(3).

Windus, Jennifer, and Marleen Kromer. 2001. *Invasive Plants of Ohio.*

Wisconsin Department of Natural Resources. 1998. *So What Should I Plant?* PUBL-WM-223–98. Madison, WI.

———. 2001. *Invasive Plant Plan Review Draft.* Madison, WI, 2001.

Wisconsin Department of Natural Resources, Bureau of Forestry. 1990. *Forest Trees of Wisconsin.*

Wisconsin Department of Natural Resources, University of Wisconsin Sea Grant Institute, the U.S. Fish and Wildlife Service, and the U.S. Army Corps of Engineers. *A Field Guide to Aquatic Exotic Plants and Animals.* Publication No. WR 407.

Wixted, Dan, Roger Flashinski, and Jerry Doll. 1998. *Training Manual for the Commercial Pesticide Applicator.* Madison: University of Wisconsin Extension, Department of Agronomy.

Woods, Brock. 2001. "Purple Loosestrife Control in Wisconsin: Status, Options, and Resources." Presentation at the Plants Out of Place Conference, Eau Claire, WI.

Zim, Herbert S., and Alexander C. Martin. 1987. *Trees.* New York: Golden Press.

Web Sites

1 Up Info. Wildlife, Animals, and Plants. http://reference.allrefer.com/wildlife-plants-animals/

Ann Arbor Natural Areas Preservation Division Invasive Species List—www.a2gov.org/Parks/NAP/AAInvasiveSpeciesList.May2004.pdf.

Aquatic Plant Management Society. "Plant Fact Sheet." www.apms.org/plants/plants.htm.

Athenic Systems. "Tree Guide, Species Details." www.treeguide.com/.

Bark, Shaun. "Giant Hogweed." The Super Byway. www.superbyway.net/hogweed.htm.

Brooklyn Botanic Garden. "The Worst Invasives in the New York Metropolitan Area." www.bbg.org/gar2/pestalerts/invasives/worst_nym.html.

———. "The Worst Invasives in the United States." www.bbg.org/gar2/pestalerts/invasives/worst_us.html.

CalFlora Taxon Report. www.calflora.org/.

Canadian Botanical Conservation Network. www.rbg.ca/cbcn/en/projects/invasives/invade1.html.

Conneticut Invasive Plant Working Group. www.hort.uconn.edu/cipwg.

Cooperative Agriculture Pest Survey Program. "Pest Information." www.ceris.purdue.edu/napis/pests/.

Cornell University Poisonous Plants Informational Database. www.ansci.cornell.edu/plants/alphalist.html.

Doll, Jerry, and Chris Boerboom. 2003. University of Wisconsin, Department of Agronomy. "Useful Plant Identification References." ipcm.wisc.edu/uw_weeds/extension/articles/plntidref.htm.

Downs, Daphne, Chris Tsai, Brett Begley, and Janet Triplett. "Poisonous Plants." University of Pennsylvania. http://cal.vet.upenn.edu/poison.

Florida Department of Environmental Protection, Bureau of Invasive Plant Management. "Weed Alert: Water-hyacinth." www.dep.state.fl.us/lands/invaspec/2ndlevpgs/pdfs/hyacinth.pdf.

Florida Exotic Pest Plant Council. "*Eichhornia crassipes.*" www.fleppc.org/pdf/Eichhornia%20crassipes.pdf.

Illinois Native Plant Society Invasive Plants List. www.ill-inps.org/index_page0005.htm.

Illinois Nature Preserves Commission. "Vegetation Management Guide." www.inhs.uiuc.edu/chf/outreach/VMG/VMG.html.

Indiana Division of Forestry, Indiana Department of Natural Resources. "Stewardship Notes, Grapevines." www.in.gov/dnr/forestry/pdfs/grapevines.pdf.

Indiana Native Plant and Wildflower Society. "Invasive Plants in Indiana." inpaws.org/plants.html.

Invasive Alien Plant Species of Virginia: www.dcr.state.va.us/dnh/invinfo.htm.

Invasive Plant Atlas of New England. http://invasives.eeb.uconn.edu/ipane.

Invasive Plant Council of New York State. "Water Chestnut—*Trapa natans.*" www.ipcnys.org/invasive_species/trapa%20_natans.htm.

Iowa Association of Naturalists. "Iowa's Shrubs and Vines." www.extension.iastate.edu/Publications/IAN307.pdf.

Ladd, D. and B. Churchwell. "Ecological and Floristic Assessment of Missouri Foundation Lands. Appendix 2: Exotic Species Table." www.moprairie.org/floristic/app2.html.

Manitoba Weeds, Insects, and Diseases: www.gov.mb.ca/agriculture/crops/wid.html.

Maryland Department of Natural Resources. "Water Chestnut." www.dnr.state.md.us/bay/sav/water_chestnut.html.

Michigan State University Extension. "Invasive Weeds of the Upper Peninsula (of Michigan)." http://forestry.msu.edu/mipc/UPweeds.htm.

Missouri Botanical Garden. "Missouri Exotic Pest Plants." www.mobot.org/MOBOT/research/mepp/

The Nature Conservancy Wildland Invasive Species Team. "Invasives on the Web." http://tncweeds.ucdavis.edu/esadocs.html.

The Nature Conservancy's Wildland Weeds Program. http://tncweeds.ucdavis.edu.

New England Wildflower Society. www.newfs.org/conserve/invasive.htm.

New Hampshire Department of Environmental Services, Water Division. Environmental Fact Sheet "Water

Chestnut Discovered in New Hampshire Waters." http://www.des.state.nh.us/factsheets/bb/bb-43.htm.

Northern Prairie Wildlife Research Center. "Species Abstracts of Highly Disruptive Exotic Plants." www.greatplains.org/resource/invasive.htm.

Ohio Perennial and Biennial Weed Guide. http://www.oardc.ohio-state.edu/weedguide/listall.asp

Ontario Noxious Weeds—www.ovma.on.ca/nox.htm.

Pennsylvania Department of Conservation and Natural Resources. "What Is an Invasive Plant?" www.dcnr.state.pa.us/forestry/wildplant/invasive.plants.aspx.

Plant Conservation Alliance, Alien Plant Working Group. "Exotic Bush Honeysuckles" www.nps.gov/plants/alien/fact/loni1.htm.

———. "Weeds Gone Wild—Alien Plant Invaders of Natural Areas." www.nps.gov/plants/alien/index.htm.

Plants for a Future. "Database Search Results." www.scs.leeds.ac.uk/pfaf/.

Purdue University Center for New Crops and Plant Products. www.hort.purdue.edu/newcrop.

Tu, M., C. Hurd, and J. M. Randall. "Weed Control Methods Handbook." The Nature Conservancy, Version: April 2001. http://tncweeds.ucdavis.edu/handbook.html.

United States Department of Agriculture, Agricultural Research Service. "Germplasm Resources Information Network (GRIN)." www.ars-grin.gov/npgs/aboutgrin.html.

United States Department of Agriculture, Forest Service. "Fire Effects Information." www.fs.fed.us/database/feis/plants.

United States Department of Agriculture, Forest Service, Eastern Region. "Noxious Weeds and Invasive Non-native Plants." www.fs.fed.us/r9/wildlife/range/weed/.

United States Department of Agriculture, Natural Resources Conservation Service. "Plants Database." http://plants.usda.gov/.

University of Florida, Center for Aquatic and Invasive Plants. http://aquat1.ifas.ufl.edu/.

University of Illinois at Urbana-Champaign, College of Agriculture, Agricultural Experiment Station. 1981. *Weeds of the North Central States.* www.ag.uiuc.edu/~vista/html_pubs/WEEDS/list.html

University of Minnesota Extension Service. "Goutweed." www.extension.umn.edu/distribution/horticulture/components/7566–49.html.

Virginia Department of Conservation and Recreation. "Invasive Alien Plant Species of Virginia—Hydrilla." www.dcr.state.va.us/dnh/fshyve.pdf.

Washington Native Plant Society. "Preliminary List of Exotic Pest Plants of Greatest Ecological Concern in Oregon and Washington." www.wnps.org/eppclist.html.

Washington State Department of Ecology. "Non-Native Freshwater Plants." www.ecy.wa.gov/programs/wq/plants/weeds/.

Westbrooks, Randy. Giant hogweed fact sheet. www.ceris.purdue.edu/napis/pests/ghw/facts.txt.

Western Aquatic Plant Management Society. "Hydrilla." www.wapms.org/plants/hydrilla.html.

Wildlife Habitat Council, Three Rivers Habitat Partnership. "Backyard Buffers." www.wildlifehc.org/threerivers/backyard_buffers/.

Wisconsin Department of Natural Resources. "Invasive Species." www.dnr.state.wi.us/org/land/er/invasive/index.htm.

———. 2003. "Herbicides for Forest Management 2003." University of Wisconsin Extension, Forestry Facts No. 76. http://forest.wisc.edu/extension/publications/76.pdf.

Wisconsin State Herbarium. http://www.botany.wisc.edu/herbarium/.

Wisconsin Vascular Plant Species. www.botany.wisc.edu/wisflora/.

Index